'What is good economics? Instead of offering armchair answers to this question, the book offers detailed descriptions of the practice of economics throughout the course of its rich history.'
            Esther-Mirjam Sent, Radboud University Nijmegen, the Netherlands

'Harro Maas provides a concise introduction to the sources of factual knowledge and approaches to theoretical knowledge of leading economists of the last two centuries. His book is not about what economists say they do, or what methodologists say they should do, but what they do. It makes for a great read.'
            Richard Van Den Berg, Kingston University London, UK

'The book's historical approach reflects a wide knowledge and deep understanding of economics and contains much that will fascinate practising economists and their students. It is a remarkable book, written in a lively and engaging style, that everyone with any interest in economic methodology, as well as economists who think methodology has nothing to say to them, should read.'
            Roger Backhouse, University of Birmingham, UK

Economic Methodology

Ever since the inception of economics over two hundred years ago, the tools at the discipline's disposal have grown more and more sophisticated. This book provides a historical introduction to the methodology of economics through the eyes of economists.

The story begins with John Stuart Mill's seminal essay from 1836 on the definition and method of political economy, which is then followed by an examination of how the actual practices of economists changed over time to such an extent that they altered not only their methods of enquiry, but also their self-perception as economists. Beginning as intellectuals and journalists operating to a large extent in the public sphere, they then transformed into experts who developed their tools of research increasingly behind the scenes. No longer did they try to influence policy agendas through public discourse; rather they targeted policymakers directly and with instruments that showed them as independent and objective policy advisors, the tools of the trade changing all the while.

In order to shed light on the evolution of economic methodology, this book takes carefully selected snapshots from the discipline's history. It tracks the process of development through the nineteenth and twentieth centuries, analysing the growth of empirical and mathematical modelling. It also looks at the emergence of the experiment in economics, in addition to the similarities and differences between modelling and experimentation.

This book will be relevant reading for students and academics in the fields of economic methodology, history of economics, and history and philosophy of the social sciences.

Harro Maas is Associate Professor in the School of Economics at Utrecht University, the Netherlands.

Economics as Social Theory
Series edited by Tony Lawson
University of Cambridge

Social theory is experiencing something of a revival within economics. Critical analyses of the particular nature of the subject matter of social studies and of the types of method, categories and modes of explanation that can legitimately be endorsed for the scientific study of social objects are re-emerging. Economists are again addressing such issues as the relationship between agency and structure, between economy and the rest of society, and between the enquirer and the object of enquiry. There is a renewed interest in elaborating basic categories such as causation, competition, culture, discrimination, evolution, money, need, order, organization, power probability, process, rationality, technology, time, truth, uncertainty, value, etc.

The objective for this series is to facilitate this revival further. In contemporary economics, the label 'theory' has been appropriated by a group that confines itself to largely asocial, ahistorical, mathematical 'modelling'. *Economics as Social Theory* thus reclaims the 'theory' label, offering a platform for alternative rigorous, but broader and more critical conceptions of theorizing.

Other titles in this series include:

1. Economics and Language
Edited by Willie Henderson

2. Rationality, Institutions and Economic Methodology
Edited by Uskali Mäki, Bo Gustafsson and Christian Knudsen

3. New Directions in Economic Methodology
Edited by Roger Backhouse

4. Who Pays for the Kids?
Nancy Folbre

5. Rules and Choice in Economics
Viktor Vanberg

6. Beyond Rhetoric and Realism in Economics
Thomas A. Boylan and Paschal F. O'Gorman

7. Feminism, Objectivity and Economics
Julie A. Nelson

8. Economic Evolution
Jack J. Vromen

9. Economics and Reality
Tony Lawson

10. The Market
John O' Neill

11. Economics and Utopia
Geoff Hodgson

12. Critical Realism in Economics
Edited by Steve Fleetwood

13. The New Economic Criticism
Edited by Martha Woodmansee and Mark Osteeen

14. What do Economists Know?
Edited by Robert F. Garnett, Jr.

15. Postmodernism, Economics and Knowledge
Edited by Stephen Cullenberg, Jack Amariglio and David F. Ruccio

16. **The Values of Economics**
An Aristotelian perspective
Irene van Staveren

17. **How Economics Forgot History**
The problem of historical specificity in social science
Geoffrey M. Hodgson

18. **Intersubjectivity in Economics**
Agents and structures
Edward Fullbrook

19. **The World of Consumption, 2nd Edition**
The material and cultural revisited
Ben Fine

20. **Reorienting Economics**
Tony Lawson

21. **Toward a Feminist Philosophy of Economics**
Edited by Drucilla K. Barker and Edith Kuiper

22. **The Crisis in Economics**
Edited by Edward Fullbrook

23. **The Philosophy of Keynes' Economics**
Probability, uncertainty and convention
Edited by Jochen Runde and Sohei Mizuhara

24. **Postcolonialism Meets Economics**
Edited by Eiman O. Zein-Elabdin and S. Charusheela

25. **The Evolution of Institutional Economics**
Agency, structure and Darwinism in American institutionalism
Geoffrey M. Hodgson

26. **Transforming Economics**
Perspectives on the Critical Realist Project
Edited by Paul Lewis

27. **New Departures in Marxian Theory**
Edited by Stephen A. Resnick and Richard D. Wolff

28. **Markets, Deliberation and Environmental Value**
John O'Neill

29. **Speaking of Economics**
How to get in the conversation
Arjo Klamer

30. **From Political Economy to Economics**
Method, the social and the historical in the evolution of economic theory
Dimitris Milonakis and Ben Fine

31. **From Economics Imperialism to Freakonomics**
The shifting boundaries between economics and other social sciences
Dimitris Milonakis and Ben Fine

32. **Development and Globalization**
A Marxian class analysis
David Ruccio

33. **Introducing Money**
Mark Peacock

34. **The Cambridge Revival of Political Economy**
Nuno Ornelas Martins

35. **Understanding Development Economics**
Its challenge to development studies
Adam Fforde

36. **Economic Methodology**
A historical introduction
Harro Maas
Translated by Liz Waters

Economic Methodology
A historical introduction

Harro Maas

Translated by Liz Waters

LONDON AND NEW YORK

First published 2014
by Routledge
2 Park Square, Milton Park, Abingdon, Oxon OX14 4RN

and by Routledge
711 Third Avenue, New York, NY 10017

Routledge is an imprint of the Taylor & Francis Group, an informa business

© 2014 Harro Maas (author)
© 2014 Liz Waters (for the translation)

The right of Harro Maas to be identified as author of this work has been asserted by him in accordance with the Copyright, Designs and Patent Act 1988.

All rights reserved. No part of this book may be reprinted or reproduced or utilised in any form or by any electronic, mechanical, or other means, now known or hereafter invented, including photocopying and recording, or in any information storage or retrieval system, without permission in writing from the publishers.

Trademark notice: Product or corporate names may be trademarks or registered trademarks, and are used only for identification and explanation without intent to infringe.

British Library Cataloguing in Publication Data
A catalogue record for this book is available from the British Library

Library of Congress Cataloging in Publication Data
Maas, Harro.
[Spelregels van economen. English]
Economic methodology: a historical introduction / Harro Maas; translated by Liz Waters.
pages cm
Includes bibliographical references and index.
1. Economics–Methodology–History. 2. Economics–History. I. Title.
HB131.M2813 2014
330–dc23
2013033789

ISBN: 978-0-415-82284-8 (hbk)
ISBN: 978-0-415-85899-1 (pbk)
ISBN: 978-0-203-79767-9 (ebk)

Typeset in Palatino
by GreenGate Publishing Services, Tonbridge, Kent

Contents

List of figures	x
Preface	xii
Acknowledgements	xiv
1 Introduction	1
2 Economics: inductive or deductive science?	8
3 Economics and statistics	20
4 Business-cycle research: the rise of modelling	38
5 John Maynard Keynes and Jan Tinbergen: the dramatist and the model-builder	60
6 Milton Friedman and the Cowles Commission for Econometric Research: structural models and 'as if' methodology	76
7 Modelling between fact and fiction: thought experiments in economics	99
8 Experimentation in economics	115
9 Simulation with models	150
10 Economics as science: the rules of the game	171
Further reading	175
Index	181

Figures

3.1	Two pages from John Elliot Cairnes' notebook about the state of the Irish economy	24
3.2	A graph by William Stanley Jevons, probably of prices and quantities	27
3.3	A diagram that shows the surface area, population and tax proceeds of the major European states in 1801	32
4.1	The Harvard barometer	39
4.2	Limnimeter and graph from Jules Marey, *La Méthode Graphique*	41
4.3	James Watt's indicator diagram	42
4.4	Two illustrations from a CBS brochure of 1939 explaining data-processing operations for import and export statistics	44
4.5	Diagrammatical representation of CBS data-processing operations on trade statistics, visualising the complete internal procedure from data-reception, to representing and archiving the original and processed results	46
4.6	Female employees enter statistical data into the Burroughs computers	47
4.7	CBS barometer	50
4.8	Tinbergen's stylized barometer	52
4.9	Tinbergen's model of the Dutch economy	53
4.10	Tinbergen's correlation diagram	54
4.11	The relationship between economic theory and statistics	55
4.12	J.J. Polak's representation of the relationship between the structure of a model and its data	58
5.1	Professor Jan Tinbergen and the staff of the department of business-cycle research and mathematical statistics, 1936	68
5.2	John Maynard Keynes in his study in Bloomsbury, 16 March 1940	74
6.1	Mitchell's specific and reference cycle chart	80
6.2	Scatter plot, representing maize production and price differences	83

6.3	Henry L. Moore's estimate of the demand function for maize	84
7.1	Fragment of text from Paul Samuelson, An Exact Consumption-Loan Model with or without the Social Contrivance of Money	102
7.2	Front cover of Simon Stevin, *Beghinselen der weeghconst* (1586)	106
7.3	What do we see, an old woman or a young woman?	112
8.1	Social Science Experimental Laboratory (SSEL) at Caltech, Passadena	127
8.2	Diagram of experimentally generated supply and demand schedules	128
8.3	Experimental market demand and supply schedules	137
8.4	Experimental market demand and supply schedules scaled up to the level of the entire market	145
9.1	Directors of the Central Bank of the Netherlands in 1958	154
9.2	Diagrammatic representation of the stages of development of the monetary model of the Central Bank of the Netherlands	157
9.3	Graph of the Adelmans' simulation of the Klein–Goldberger model without the application of a 'shock'	159
9.4	Graph of the Adelmans' simulation of the Klein–Goldberger model after the application of random shocks	160
9.5	Diagram showing the relationship, as seen by the CPB, between expert knowledge and the model	168

Preface

This book is a historical introduction to the methodology of economics, with a strong emphasis on its practice. Those familiarizing themselves with this field of study will find little in the way of insight into how economists actually go about their work, irrespective of whether they consult introductions to the philosophy of science or textbooks about economics. A university student may well write an undergraduate thesis nowadays without any knowledge at all of the dirty manual labour that research involves. The same applies, unfortunately, to introductions to the philosophy of economics, where the chaotic day-to-day business of research – referred to so aptly by sociologist of science Andrew Pickering as the 'mangle of practice' – is tidied away into neat definitions and refined analytical problems. It is this lacuna that my book is intended to fill.

I do not mean to suggest that I have written an introduction to the methods and techniques of research in the social sciences. This is not a guide to statistical methods, econometric techniques or qualitative forms of investigation. Rather it takes the reader through the history of the subject, using examples to show how the practice of economics has developed over time. It is therefore a reflection of the courses I have taught over many years at the University of Amsterdam and subsequently at Utrecht University, along with my colleagues the late Mark Blaug, Marcel Boumans, John Davis, Floris Heukelom, Mary Morgan, Geert Reuten, Peter Rodenburg and Tiago Mata. Without my students and my regular exchanges of ideas with fellow academics, it could not have been written. I thank them all sincerely, including those at the economics faculty at Amsterdam and elsewhere who offered me insights into their working methods. I would like to thank Charlie Plott, Andrej Svorencik and Frans van Winden in particular for in-depth discussions on experimental economics. The chapter about simulations could not have been written without close contact with Frank den Butter, and I thank him especially for giving me access to his personal archive. Conversations with Martin Fase, Wim Vanthoor, Johan Verbruggen and Henk Kranendonk were especially helpful for that particular chapter.

Geerte Wachter, Floris Heukelom, Peter Rodenburg and two editors of the series in which the Dutch edition of this book appeared in 2010, Rob van Es and Chunglin Kwa, read the entire manuscript. I am extremely grateful to all of them, especially Floris and Peter. A short stay at the Max Planck Institute for the History of Science in Berlin in spring 2010 gave me precisely the boost I needed to complete the book.

For the English edition, I drew upon my 'Questions of scale in economic laboratory experiments', published in *Revue de Philosophie Economique*, 13(1) (2012). I also reconsidered materials in some of the other chapters, especially Chapter 9. Section 3.2 is loosely based on 'Sorting Things Out: The Economist as an Armchair Observer', published in Lorraine Daston and Elizabeth Lunbeck (eds) *Histories of Scientific Observation* (University of Chicago Press, 2011), pp. 206–229. I would like to thank Liz Waters for clarifying my thoughts, and Robert Langham and Natalie Tomlinson for their support in producing this book. This English version is in memory of Mark Blaug who, I am sure, would have disagreed with its substance and would not have spared me his astute comments.

Harro Maas
Amsterdam, 1 April 2013

Acknowledgements

The author and publisher greatly acknowledge the generous support from the Netherlands Foundation for Scientific Research, grant 276-53-004, and the Dutch Foundation for Literature in the production of this book.

Nederlands letterenfonds
dutch foundation for literature

1 Introduction

In the 1966 film *Blow-Up* by Michelangelo Antonioni, a young photographer accidentally and unknowingly witnesses a murder. Taking pictures in a quiet park in the swinging London of the 1960s, he follows a couple, a woman and a rather older man, who are obviously in love. When the woman becomes aware of the photographer, she runs after him and demands his roll of film. The photographer refuses to give it to her, but on arriving home he finds she has followed him, still insisting he must hand over all the negatives. The photographer lets the woman in and, although he plays it tough, he promises to give her the film. By this point, he is so intrigued by her persistence that rather than letting her have the original roll, he switches it for another. After the woman has left, he develops the photographs in large format and hangs them up side by side in his studio. Only then does he notice that the woman looks tense, staring into the bushes. He marks the precise place she is looking at, blows it up (hence the name of the film) and sees that someone appears to be hiding in the shrubbery. Then he spots a revolver in the half-concealed person's hand. Has he witnessed a murder? When the photographer returns to the park at night to check, there is indeed a body, but back in his studio he discovers that the prints and negatives have all been stolen. The next day there is no longer any trace of a corpse. What did he actually see? Has he simply imagined the whole thing?

The way the photographer in the film goes about his work is analogous to the method used by a scientist, including the kind of scientist this book is about, the economist. Driven by curiosity as to why the woman attaches such value to his photographs (what, in other words, their compromising character might be), the photographer deploys all the techniques and equipment at his disposal, determined to find out more. Investigating the clues in the material, he discovers he has actually witnessed a situation quite different from what he thought he was seeing. He discovers this not in the field but back at his studio, where he has a darkroom, lighting equipment and cameras available for use in puzzling out, in peace and quiet, what happened in the field. At this stage, he does not see the body directly, but his equipment enables him to produce a photograph of

2 *Introduction*

a photograph. Rather than direct observation, indirect observation, aided by instruments, leads him to the astounding conclusion that he may have been present at a killing. He draws this conclusion at home, well away from the scene, the way Sherlock Holmes might, although in this case based on photos, or on photos of photos. The original photograph serves as material or data for examination back at his studio. The confirmation he seeks in the park at night is worth little if he cannot corroborate it with evidence both accessible to and verifiable by a third party.

The photographer has a darkroom and all kinds of appliances that he can use to reveal unexpected evidence, just as the scientist has a dedicated workspace and apparatus with which to make his sometimes surprising observations about the world. It is after all unusual nowadays for a scientist to study nature in situ, in the field. This book is about the techniques and instruments that economists have developed to enable them to reveal connections in a complex reality in which they would otherwise remain hidden.

The techniques of the political economist were traditionally of a different character from those of the natural scientist. True, the natural sciences were held out to the economist from the start as an example to follow, but always on the understanding that he would inevitably have to manage without the instruments available to the physicist. Laboratories, experimentation and precise, scientifically quantified laws lay beyond the reach of the science of economics. The socio-economic reality was too complex, the factors that influence human behaviour too diverse, to allow for meticulous experimentation in laboratory conditions.

Instead of controlled experiments, economists have used all kinds of other techniques to help them comprehend economic reality and, where possible, to intervene in it. In the nineteenth century, the techniques of political economy consisted of combinatorial skills, plus pen and paper – the very techniques that historians relied upon. Using his own powers of judgement, the political economist combined his various sources of information to create a coherent picture, which was then placed at the service of political intervention in the public domain. In Britain especially, it was a short step from writing an article or book to raising questions in parliament. David Ricardo, one of the most important political economists of the early nineteenth century, was himself a Member of Parliament. John Stuart Mill, an enthusiastic proponent of Ricardo's theories, was the editor of *The Westminster Review*, a magazine by and for political radicals in the House of Commons.

Not until the twentieth century did these techniques assume a form that economists, or the academics among them at least, took to be self-evident. Some experienced as a liberation the fact that not just economists but other scientists too, including meteorologists and astronomers, had come up against the limitations of the controlled experiment. All of them, economists included, now construct mathematical models that use mathematics and statistics to create a picture of reality. Economists use data sets from statistical bureaus (the CBS in the Netherlands, the Bureau of Economic Analysis in the US, the Office for National Statistics in the UK and so on) to compile visual representations of socio-economic reality that cannot be seen with the naked eye.

The introduction of the modelling approach into economics changed both the relationship between economics and the natural sciences and the relationship between political economists and the public domain. Towards the end of the nineteenth century, economists began to call themselves just that, rather than 'political economists'. They became wary of intervening too conspicuously in the public sphere. Although they continued to write newspaper columns and letters to the editor, such efforts were no longer regarded as the very essence of their work. The appropriate outlets for their writing were more specialized. New periodicals emerged to publish the work from which economists derive their status, which can now be found even in publications that focus on pure science, such as *Nature* and *Science*.

As a result, economists like today's Paul Krugman, whose influence stems to a great degree from his column in *The New York Times* rather than directly from his status as a winner of the Nobel Prize for Economics, create an almost tangible sense of unease among academics, since they do not fit the cherished image of the economist as a pure scientist who writes difficult mathematical articles for important but little-read international journals. In the first decade of the twentieth century, when economics as a discipline adopted the language of mathematics and identified quantitative statistical facts as the only data admissible in empirical research, the economist seemed to have thrown off for good the image (particularly prevalent in Britain) of public intellectual and critical interventionist in the public debate, an image that had been so pronounced in the case of John Stuart Mill or John Maynard Keynes. The 'television economists' we see and read of today are regarded with a mixture of scorn and envy for having abandoned the truly scientific work.

It is no coincidence that one of the founders of social statistics, a Belgian called Adolphe Quételet (1796–1874), was director of the astronomical observatory in Brussels. With his arrival on the scene, the character of the economist, his 'persona' as it is called nowadays in the history of science, changed to become a hybrid between natural scientist and social scientist. Quételet described his use of social statistics as *physique sociale*, as if he had found a way to train his telescope on society. Just as observatories were built higher and higher in the mountains, so the economist withdrew from the front line of public debate and was able to use mathematical models, supported by statistics, to influence socio-economic policy at least as effectively as he had done in earlier years.

What are the rules that economists abide by in their research and how do those rules of the game determine the character of a good economist? These questions recall a book by Joop Klant, a historian and methodologist of economics, that was published in Dutch over forty years ago with the title *Spelregels voor economen* (Rules of the Game for Economists, 1972). Klant was writing in a period when the philosophy of science was promising to provide rules for economists to adhere to in their actual research. There was an expectation that prescriptive or normative methodology, as it was known, would improve the quality of economic studies. Klant's book was extremely readable and informative, applying to economics the philosophy of science as it then stood, but it did not explain why economists, who after all are pretty smart people, use different rules in practice from those he advocated. Klant's *Spelregels* defined how economics must be practised if it is to be and remain a sound science. It also defined the persona of the economist: an academic who tests theories as part of a quest for objective knowledge. The Dutch edition of the book you are now reading was published with a title intended as a pun on Klant's, namely *Spelregels van econtomen* (*Economists' Rules of the Game*). It seeks to explain the actual rules as applied by economists now and in the past, and it was written in the conviction that rules governing research are not timeless but historical, developed over many years and continuing to develop. To borrow from the title of Bruno Latour's landmark work, *Science in Action* (1987), this is a book about economists in action.

So here you will read about changes in the methods and techniques that economists have used to analyse reality, and about the accompanying changes in the nature of the scientific and public role of the economist. I aim to show both sides of the coin. By examining the history of economics, we can trace changes both to the instruments of economic science and to the character of the economist as a scientist. Which methods are right and what, therefore, constitutes good economics? These are not established matters of fact. My approach, taking its lead from the history of science, might be described as historical epistemology. I do not attempt to provide rules for economists but instead to describe the rules by which they have been governed and which they have developed for themselves.

The idea that an economist is like a photographer brings to mind a book by Donald MacKenzie published in 2006, in which he analyses the origins and development of the economics of the financial markets: *An Engine, Not a Camera*. The title is taken from a remark by economist Alfred Marshall about the nature of economic models. In the 1950s, Chicago economist Milton Friedman used the same notion to defend his own concept of economic science, claiming that Marshall believed such models functioned as engines rather than cameras in that they did not depict reality but were instruments that could be used to change it.

The modelling approach is unquestionably dominant in contemporary economics, so an important part of this book is about the origins and development of modelling as a twentieth-century research instrument, but we cannot be certain that Friedman is right to claim that models are machines for investigating and changing reality, rather than ways of representing it. The example of the film *Blow-Up* suggests that models are in fact both instruments and representations.

Does an economic model indeed function in the same way as a photographer's camera or an astronomer's telescope? Or is it the other way around? Should we see such a model as the real economy on a smaller scale? If such a comparison holds true, then where is the economist's dark room? How and where are the pictures developed? How can he know that the image he has created by using his instruments does in fact reflect reality?

These are the questions I address in this book. What kind of instruments do economists use to enable them to understand and intervene in the world? How have these instruments developed over time? How have changes to the available instruments changed the persona of the economist? How in turn has the social playing field changed, in other words the terrain on which the economist moves?

As I have already indicated, my approach to these questions is a historical one. I use a number of historical snapshots or case studies to build up a picture of the way in which the methods used by economists have developed over time, and to make clear the problems and opportunities they encountered along the way. I do not intend to offer a linear history with the current situation as its triumphant culmination, although clearly an approach using mathematical models dominates economics as currently practised. With the exception of a short outing to the German language area in Chapter 3, I take my examples from the Anglo-Saxon countries and my home nation, the Netherlands. This is a deliberate choice, made in the light of the specific themes I have chosen to address in the book.

In Chapter 2, I look at an essay by nineteenth-century British political economist and philosopher John Stuart Mill about the definition of political economy and its methods. Mill's essay is still an important reference point in discussions about methodology. He highlights three themes that are important throughout this book: the complexity of economic reality; the specific, indeed unique character of economics as a science; and the role of the economist in society. We will look at how economists have attempted to deal with the problem of complexity, examining the implications of those efforts for the relationship between economics and other disciplines, and for the profile of the economist as a scientist.

Chapter 3 shows how any shift in the nature of the instruments used by an economist to tackle complex reality changes the persona of the economist himself. He is transformed from a public debater into a hybrid expert, a combination of engineer and social scientist, working behind the scenes to develop ways of creating as objective a picture of economic reality as possible. At one time a science among other social sciences, economics has become a branch of learning that shares with disciplines such as astronomy and meteorology a particular stress on mathematics and statistics.

The mathematical and statistical approach to economics began to dominate the discipline from the Second World War onwards, and in

Chapters 4 and 5 I look at what this meant in practice. I describe the rise of economic modelling and show how the philosophical approach seen in Keynes and his predecessors gave way to an approach in which mathematics and statistics were used to build miniature versions of real-world economies, known as models.

Chapters 6–9 all deal in one way or another with models. What kind of instruments are they? Do they serve to predict the course of events without any need for us to worry about the specific structure of the model itself, or is the whole point of models that they offer us insight into the mechanisms that we can justifiably assume to be operating in reality? These chapters fall somewhere between the radical position of Milton Friedman (Chapter 6), who argues that the assumptions at the root of models are irrelevant, and positions that in one way or another support the notion that those assumptions are true, as in Chapter 9. In the former case, the value of models lies primarily in their predictive power, in the latter case in their capacity to simulate reality.

Chapter 7 looks at the work of Paul Samuelson, showing that models can also facilitate thought experiments. In contrast to modelling exercises that link model and world directly, thought experiments hinge on paradoxes that undermine common tenets of our understanding of the economy. One might say that thought experiments test common sense and therefore indirectly show how the world functions differently from the way we believe it does. Chapter 8 is about material experimentation in economics. It describes a fundamental shift in both thought and action, the consensus among economists having previously been that their particular science did not lend itself to experimentation. We shall see how, since the 1960s, economists have laid aside a number of standard arguments against the feasibility of experimentation, not simply by arguing that it is feasible but by actually going ahead and experimenting.

With the exception of Chapter 2, which looks at John Stuart Mill's essay, each of the chapters addresses one or more examples of economic research, and these in turn dictate the subjects treated. To that extent, this book differs from other introductions to the philosophy and methodology of economics, which take their lead from the philosophy of science. I certainly do not intend to suggest this tradition is irrelevant to attempts to understand economics, but it is not my aim to show how the classic concerns of the philosophy of science arise when we think about economics as a science. Rather, I intend to demonstrate how economists themselves have expressed and attempted to solve the problems they come up against in practice. The book can therefore be read alongside either a more traditional introduction to the philosophy of science or an introduction to economics. Of course, it can also simply be read for its own sake.

In the concluding chapter, I return to the character of economics as a science and what it implies for the persona of the economist. A historical approach to the methodology of economics demonstrates that its rules cannot be expressed as timeless truths. Nor is the persona of the economist as a scientist stable and fixed. Both the rules of the game and the character of its players change as the context and methods of economics change. Although still a participant in public debates, to an important degree the economist has become an instrument maker. Yet despite this, he remains a player. Indeed, his opinion may actually have a greater influence on public policy as a result.

2 Economics
Inductive or deductive science?

2.1 Introduction

In 1836, British political economist John Stuart Mill (1806–73) published an essay entitled 'On the Definition of Political Economy; and on the Method of Philosophical Investigation in that Science' in *The Westminster Review*, the magazine of Britain's radical political reformers. His analysis enabled economics, as a science, to rise above the social and political battle of which it was part. Mill was certainly not alone in undertaking this task, but his attempt was the most successful. Along with the final volume of his *A System of Logic*, published in 1843, which looks at the methods of what were then called the 'moral sciences', his essay continues to shape our thinking about the methods of economics. This chapter looks at Mill's essay by placing it in its historical context. In subsequent chapters, we will examine more closely the effects of the essay over time, and see how economists have used Mill's ideas to express, and to a certain degree to solve, the methodological problems they face.

2.2 Mill's life and work

John Stuart Mill was born in London in 1806 and died in 1873 near Avignon, where he is buried. Throughout his life, he found himself at the heart of the political controversies of Victorian England. In his youth, he was arrested and held for several days for handing out condoms in the street, at a time when birth control was regarded as a deeply vulgar business. For many years, he had a platonic relationship with feminist Harriet Taylor, whom he married when her husband died. As a rather less than successful member of the British House of Commons, he attempted to break the power of the landed aristocracy and, in his later years, he showed some cautious sympathy for socialism, an ideology still in its infancy. But Mill has gone down in history above all as one of the founders of classical liberalism and as one of the most important British philosophers and political economists of the nineteenth century.

John Stuart Mill was the son of James Mill (1773–1836), who moved from Scotland to London at the turn of the nineteenth century, when still in his twenties, to make a living as a journalist. It was an extraordinary step in those days. James Mill quickly became friends with two men who likewise left their mark on their era: Jeremy Bentham (1748–1832), a radical political reformer who helped to write the new French constitution after the French Revolution of 1789, and David Ricardo (1772–1832), a stock trader, major landowner and Member of Parliament, with roots in Jewish Amsterdam. We remember Bentham as the founder of utilitarianism in ethics, which he described by using the famous phrase 'the greatest happiness for the greatest number'. David Ricardo is among the most important of classic political economists, alongside Adam Smith (1723–90), Robert Malthus (1766–1834) and Karl Marx (1818–83).

From early childhood, Mill was raised and educated entirely by his father. In his autobiography (which should be read with a degree of scepticism), he describes in detail the extraordinary circumstances of those early years. By the age of about four, he could read Latin and Greek; by the time he turned twelve, he had read virtually all the important works of classical antiquity. In his early teens, he served as personal secretary to his father and to Jeremy Bentham, so he was familiar with the most minute details of their ideas even prior to publication. He became editor of *The Westminster Review*, a new periodical that provided a platform to radical social reformers such as Bentham and James Mill. Just as we cannot understand our own times without reference to the internet, Victorian England is incomprehensible without taking into account the roughly 125,000 newspapers and magazines that circulated for long or short periods over the course of the nineteenth century. Opinions were expressed publicly not through blogs, Twitter and YouTube but by means of the pen and the printing press.

2.3 The background to Mill's *Essay*: Malthus and Ricardo

In the early nineteenth century, political economy had a pivotal place in public debate in Britain. Discussion was dominated by rapid increases in the price of grain as a result of the Napoleonic Wars, fears of a population explosion, famine and impoverishment among large sectors of the population and the radicalization of the populace. Of particular concern were the population issue, agricultural productivity and, closely related to both, the position of the aristocracy. Robert Malthus, who from 1805 onwards occupied Britain's first professional chair in political economy, believed that population growth was a huge and increasing problem that could be tackled only by abolishing the Poor Laws that had governed support for the poor for centuries and were first systematized in 1601 under Queen Elizabeth I. It seemed that the Poor Laws, despite a series of amendments over the years, had succeeded only in increasing the number of those unable to fend for themselves.

At Jesus College, Cambridge, Malthus had trained for the Anglican ministry and indeed he served for some ten years as a country curate in Oakwood, Surrey. It was there that he published his renowned *An Essay on the Principle of Population* (1798). In it he takes on Enlightenment thinker William Godwin, who had written a hugely popular book describing a utopian vision of a society based not on self-interest and property but on altruism and free love. Malthus uses a thought experiment to show that Godwin's perfect society could never last but would quickly revert to our own imperfect order. At a stroke he put paid to the eighteenth-century Enlightenment's promise of a perfectly rational world and of prosperity for all.

Malthus based his ideas, in part at least, on the growth of the population in the new settlements in North America, where there was no shortage of land, and on his own experiences at Oakwood. His successors were astonished by the rise in population recorded in village registers of births and deaths. To Malthus, this was crucial evidence for his conjecture that the growth of food production could not keep pace with population growth.

Ricardo was another British intellectual who saw population growth, combined with a limited ability to supply everyone with sufficient food, as a problem of central importance. He believed it was obvious that the best agricultural land would be worked first and therefore that land farmed later would be less productive. New agricultural land, Ricardo (like Malthus) believed, was characterized by what we would now call diminishing returns.

Ricardo argued that this meant landowners were able to put a price on better land, demanding ground rent for it. There was no need for them to do anything; it was a natural consequence of the pressure of population and the expansion of farmland that resulted. Call it 'unearned income'. Aided by simple calculations, sometimes included in the text of his *On the Principles of Political Economy and Taxation* (1817), sometimes in separate tables, Ricardo developed various scenarios that demonstrated the results of steadily rising land rents for the incomes of the other two social classes in the economy: the capitalists and the workers. The workers were reduced to subsistence level and the income of the capitalists (in other words their profits) eventually declined to zero.

Ricardo makes clear that aristocratic wealth is based on purely accidental circumstances and gained at the expense of the other classes in society. The political conclusion was quickly drawn: the enrichment of the landowning classes was holding back the growth of rapidly developing industries in cities such as Manchester and Liverpool. Although Ricardo was initially interested in the countryside (it was only in the third edition of 1821 that he added his famous chapter on machinery), his theory clearly seemed to imply that capitalists, the entrepreneurs setting up new industries and taking on fresh challenges, were the engine of progress in England, rather than the aristocracy or the church.

These political conclusions were of course unacceptable to the establishment. Practically all the land in England was owned by the aristocracy and the Anglican Church, and the story went that you could walk from Oxford to Cambridge on land owned by the colleges. The famous struggle in the first half of the nineteenth century over the abolition of the Corn Laws, which were keeping the price of grain artificially high, was directly connected with the position of the aristocracy, the church and the impoverished rural population. So soon after the French Revolution and the period of the Terror, and with the destruction that Napoleon had wrought upon Europe still fresh in everyone's mind, the British establishment was not exactly eager to hear Ricardo's revolutionary message.

2.4 Reaction of the establishment

The establishment's anxiety was wonderfully portrayed in the literature of the time. English Romantic poet William Wordsworth, for example, who studied at St John's College, Cambridge, focuses in a number of his poems on the 'false philosophy' of the political economists and (in a reference to Adam Smith) the 'utter hollowness of what we name the Wealth of Nations' (Wordsworth, 1805). A few decades later, Charles Dickens created immortal characters such as Gradgrind and Scrooge (the Gordon Gekko of his day) in *Hard Times* and *A Christmas Carol* as a means of accusing the political economists of being interested only in acquisition and in 'facts, facts, facts'. In a fierce polemic aimed at John Stuart Mill (on the subject of the abolition of slavery), keen-witted conservative commentator Thomas Carlyle described political economy as 'the dismal science'.

The battle against political economy was fought outside literary circles as well. Philosophers and scientists at Oxford and Cambridge attempted to detach it from the political radicalism of Ricardo. In Oxford, Nassau Senior and Richard Whately were the main figures attempting to neutralize Ricardo's explosive message, in order to defend the interests of the Anglican Church. In early 1830, Senior was an important member of the commission charged with revising the Poor Laws and, in 1832, Whately became Anglican Archbishop of Dublin. In Cambridge, William Whewell and his good friend Richard Jones were the most prominent voices in the debate over the theory and methods of political economists. Jones succeeded Malthus as professor at the East India College in the 1830s, while as Master of Trinity College, Cambridge, Whewell became perhaps the most important spider in the web of Victorian science. As one of the founders of the British Association for the Advancement of Science (BAAS), he helped to institutionalize the 'rules of the game' that governed the way scientists engaged in reasoned debate. His *History of the Inductive Sciences* and *Philosophy of the Inductive Sciences*, published in 1837 and 1840 respectively, have ever since been treasure troves for historians and philosophers of science. It was Whewell who coined the term 'scientist'.

2.5 Whewell's insistence that political economy is an inductive science

It has been said of Whewell (son of a Lake District carpenter) that he regarded only two things as truly important: Trinity College and the Anglican Church. He therefore watched with horror the rise of political radicalism that threatened to undermine both large landowners and the church. He attacked the claims to knowledge of Ricardo's discipline, specifically the scientific methods the political economists deployed. At the heart of the dispute were the status and consequences of the presumption of self-interest.

Whewell believed that political economists such as Ricardo were wrong to generalize by elevating to the status of a universal law the 'trivial fact' that some people allowed themselves to be led by self-interest. Instead, they would do better to follow the inductive route advocated by Francis Bacon in the seventeenth century, which suggested that all science ought to begin with a careful and comprehensive description of the facts. Those facts must then be classified and labelled, and only then was it possible to look for regularities and perhaps laws. The great mistake of the 'rotten, pseudo-political economists' (Whewell, quoted in Maas, 2005) was that they extrapolated general laws without a proper investigation of the facts of the matter, with the result that they were 'driving tandem with one jack-ass before the other' (ibid.). Whewell was convinced that a careful investigation of the facts would show that no conflict of interest existed between the different classes in society, and that the population issue was not as dramatic in character as Malthus and Ricardo claimed.

Clever organizer that he was, Whewell tried to pull the rug out from under political radicalism not just through criticism of this sort but by institutionalizing the kind of practice of economics that he favoured. To that end, he made use of the BAAS, which he had helped to found in 1831. The explicit aim of the association was to enable a broad public to become familiar with the fruits of science and, at the same time, to show how true scientists conversed with each other, encouraging society at all levels to follow their example. As if to emphasize the unity of Britain as a whole, and to play down any conflicts of interest within it, the first meetings of the BAAS were held in Cambridge, Edinburgh, Glasgow, Dublin and Bristol, and in subsequent years the association entered the heartlands of the industrial revolution: Birmingham, Manchester and Liverpool.

In 1833, Whewell issued an invitation to the famous Belgian astronomer and statistician Adolphe Quételet, with the intention of ensuring that the BAAS would set up a department specifically devoted to economics and statistics, to be known as Section F. The following year, along with Richard Jones, computer pioneer Charles Babbage and others, Whewell was instrumental in founding the Statistical Society of London (renamed the Royal Statistical Society in 1884). It caught the attention of the establishment right from the start.

2.6 Senior and Whately: political economy as a deductive science

The shock was great, therefore, when Whewell and Jones discovered that Oxford dons Nassau Senior and Richard Whately were trying to strip Ricardo's political economy of its radical cast in a different way altogether. Whately and Senior were fellows of Oxford colleges Oriel and Magdalen and, just like their Cambridge counterparts, they were convinced that political economy represented too great a threat to the Anglican Church to be left to the radicals. Instead of advocating, like Whewell, the inductive method, they defended the deductive approach.

Whately and Senior's reasoning was as follows. There was nothing wrong with starting out from the idea that people were motivated by self-interest, but that did not mean political economists were justified in attaching political consequences to their theories. On the contrary, it was their job to make a neutral and objective study of how people acted in the market, just as natural scientists made a neutral and objective study of the laws of nature, no more and no less. It was then up to political institutions to extrapolate from the results of such studies. In other words, political economy was a positive rather than a normative discipline. Senior and Whately were attempting to defend existing social institutions (in their case primarily the Anglican Church) against the blasphemous political proposals of radicals such as Bentham, James Mill and Ricardo.

According to Whately and Senior, political economy had no option but to take a deductive approach, since its first principles were actually astonishingly simple and known to everyone. Just as nobody could avoid being a logician in everyday life, nobody could avoid knowing the laws of supply and demand from daily experience. Instead of collecting an absurd number of historical facts, as the Cambridge men were advocating, it was better to have faith in the certainty provided by such commonplace experience. The only problem that remained to dog political economy was that so many other factors influenced market results. Political economists could really only deduce relationships that, although they resembled laws, were in fact tendencies. These came to be known as 'tendency laws'. For this reason too, it was not their task to make firm political recommendations.

For Whewell and his friend Jones, this line of reasoning was utterly unacceptable. In a letter to Whewell, Jones complained that Senior and Whately were trying to force their modernist views down the throats of the British, irrespective of actual social conditions. There was a grain of truth in that. When widespread famine hit Ireland in 1846 and 1847 as a result of potato blight, killing at least a million people, Whately responded by founding the Dublin Statistical Society, whose aim was to defend political economy against unjustified attacks. Even well into the next decade, students at Trinity College Dublin were asked in their exams to describe how grain merchants, rather than driving up the price of grain disproportionately, had fulfilled their natural task in a market economy.

2.7 Mill's *Essay* and the method of political economy

We have already looked at the background to Mill's 1836 essay in *The Westminster Review* and noted that, in 1833, Whewell tried to institutionalize an alternative, inductive form of economics in Britain. It had seemed a good moment to strike the final death blow to the political economy of the Ricardians. Even in the Political Economy Club in London, where theoretical and practical issues of an economic nature were discussed monthly following a good dinner, the prestige of Ricardo's ideas was declining. When discussing the merits of Ricardian economics there in 1831, political economist Robert Torrens seems to have held the opinion that 'all the great principles of the Ricardian system had been abandoned', especially the critical ones concerning value, rent and profits (quoted in Meek, 1950).

With the establishment of Section F at the BAAS and the founding of the Statistical Society of London, Whewell seemed accurately to have sensed the turn of the tide and created the preconditions for an inductive approach to political economy that began with the gathering of empirical facts. Even though the Oxford economists, just like the Ricardians, advocated a deductive approach, they agreed with Cambridge that the radical political consequences of Ricardo's economic theory ought to be seen as external to the realm of science.

One question that will occupy our thoughts in this book is precisely what we ought to conceive as being 'empirical facts'. In providing examples in his notes on the method of political economy, Whewell names such diverse sources as quantitative statistics, conversations with farmers and entrepreneurs, and books, whether contemporary or historical. Today's economists would regard only the first of these as furnishing empirical facts.

In early 1830, John Stuart Mill began work on what would become his *A System of Logic* (1843). In the same period, he wrote his essay 'On the Definition of Political Economy', which was published in 1836 in *The Westminster Review*. In *Logic*, he quotes at length from his essay in discussing the special status of political economy among the 'moral sciences'. The essay appeared at a crucial moment, when Ricardo's political economy was facing opposition both for its content and for its methodology. In his defence of Ricardo, Mill concentrates on two issues: the meaning of and relationship between induction and deduction, and the defining characteristics of political economy as a science. Mill expected his essay to become a classic, and in that he was certainly right. As will become clear, it serves to this day as a starting point for philosophical discussion of the unique nature of economics as a science.

2.8 Mill's objections to inductive statistics

We have already seen that men such as Whewell were trying to develop an inductive alternative to Ricardo's political economy by combining diverse sources of information including quantitative statistics. Ever since Adam Smith, political economists had been distrustful of the use of statistics in economics (a distrust that was shared in Oxford). They believed that 'political arithmeticians' or 'statists', as statisticians were then called, arrived at mistaken policy recommendations based on a random collection of facts, advocating protectionism in place of free trade. This had to do with the traditional link between statistics and the interests of the state.

The use of the word 'statists' reflects this association with the state, and in many cases political arithmeticians and statisticians did indeed collect data on behalf of states, although in this sense Britain differs from continental Europe. (In Britain, statistics were often gathered on the initiative of ordinary citizens rather than the government.) The word 'statistics' had a far broader meaning than it has today. Whereas we now think purely of quantitative data, statistics once referred to all kinds of facts, quantitative or qualitative, that might be of use to the state.

In continental Europe in particular, the collection of statistical facts had taken off in the eighteenth century. In German-speaking countries, what was known as *Kameralwissenschaft* (commonly translated as cameralism and seen as the forerunner of the science of public administration) had gradually developed into an academic discipline that, among other things, charted the relative power of states. Professors at various German universities taught a subject we would now call statistics. In France, the discipline had emerged even before the Revolution, but under Napoleon it developed into an important instrument for use by the state. The gathering of statistical data was in many cases intended to give the state more control over society. This was true of England's famous Domesday Book, produced for William the Conqueror as an inventory of the population and wealth of England and parts of Wales. The same went for the Down Survey in Ireland, compiled after the Irish Rebellion of 1641 on the orders of William Petty, forefather of the system of national income accounting, to enable the English to govern (or indeed oppress) the Irish more effectively. Politically speaking, therefore, statistics and an inductive approach to political economy were by their very nature suspect to a liberal thinker such as Mill.

But Mill also had a methodological objection to the use of statistical data as a starting point in economics. He believed social reality was so complex that it was hard to see how you could distil an unambiguous link between cause and effect out of statistical data (in which that immense complexity manifested itself). Political economy was therefore in the same position as the study of history. Mill believed the job of the historian was to describe events that took place in the past and he was convinced that such work could not produce a system of laws in the way that the natural sciences, such as mechanics, could. Whereas Whewell sought ways to discover regularities or even laws in statistical data, Mill regarded that task as simply impossible, given the problem of complexity.

At the same time, it was precisely the problem of complexity that made it impossible for political economists, in contrast to natural scientists, to isolate causal links by performing experiments. How could the causal factors that influence social reality be dealt with in such a fashion that it was possible to speak of a controlled experiment? Mill regarded astronomy as another example of a science that was unable to take an experimental approach, yet in his view it was far less complex; the movement of the planets could be explained and calculated based on the laws of mechanics.

2.9 The unique character of economics as a science

If political economy, because of the problem of complexity, found itself in the same position as history, was there perhaps some means other than induction by which it could identify universal law-like regularities on a par with the laws of the natural sciences? Contemporary philosopher of science Daniel Hausman (1992) has characterized Mill's solution to this problem as the quest to define economics as an inexact yet nevertheless separate science. In other words, complexity was a fundamental obstacle, preventing political economy from arriving at reliable, rule-governed knowledge, but it was in this very problem that Mill sought a solution.

His approach was as follows. First, he established a boundary between on the one hand the 'moral sciences', which is to say the social sciences including psychology, and on the other the natural sciences, by making a categorical distinction between their respective objects of study. This distinction then provided Mill with a way of solving the problem of complexity: the 'moral sciences' took a different approach from the natural sciences to the phenomena they studied.

With these two distinctions in hand, Mill was able to argue that political economy was in a position to discover laws that were as certain as those of the natural sciences, even though it would never be possible to observe those laws in a pure form in social reality. Political economy was a science of tendencies, and those tendencies could be identified, with certainty, by means of introspection. What did Mill mean by this?

2.10 Political economy: a science of people and markets

In the first part of 'On the Definition of Political Economy', Mill discusses several different definitions, deploying them as a means of making clear that the political economist asks a different type of question about the world from that asked by the natural scientist. Take the steam engine as used in industry. A natural scientist might be interested in the principles underlying the workings of a steam engine (just as today he would be interested in the electronics used in computing). But while industry could not function without such machines (and therefore the laws and principles involved), the political economist was not interested in laws of that kind. What mattered instead were the goals someone wanted to achieve with the help of such technology. In other words, the political economist was looking for motives, which meant that the type of law he was trying to track down was in essence psychological. Mill insisted that only psychological laws, not laws of any other kind, belonged to the discipline of economics.

How was it possible to know about such motives for action and then to work out how someone would actually act based on those motives? As we have seen, concrete situations were always far too complex for anyone to say anything well-founded about them. Mill's answer was astonishingly simple and extremely convincing. He drew a comparison with the experimental scientist. One of the fundamental problems faced by the social scientist was that complex social reality made it impossible to isolate straightforward cause-and-effect relationships experimentally, but he could engage in mental experiments. After all, everyone could work out, in the private laboratory of his own mind, how he would act in a concrete situation based on specific motives, and then imagine how others would act if they found themselves in a comparable position. (Note that I do not refer to this as a 'thought experiment'. As we shall see in Chapter 7, a thought experiment may take place in the mind, but its structure is rather different.) So in the 'moral sciences', introspection was an alternative to the experimental method used in the natural sciences as a way of achieving certainty. Introspection, or experimentation in your own mind, was just as good a way to discover laws in political economy as observation and experiment were in the natural sciences.

Mill added that for the political economist, not all psychological principles were important, only those that expressed themselves in specific circumstances: decreasing marginal returns and increasing pressure of population. These were precisely the principles at the root of Ricardo's political economy, and they related to circumstances in which only a small number of motives were relevant. Furthermore, those motives counteracted each other to some degree, since the desire for prosperity was set against an aversion to work and an eagerness to consume as many goods as possible – including luxuries – immediately. The political economist, in short, looked at just one aspect of the entire social reality and examined only those motives that were relevant to that aspect. He always looked at an abstract situation, not at the concrete and complex reality.

Such situations could and inevitably would be disrupted in the real world by any number of incidental factors. Nevertheless, a political economist could make general statements about the consequences of an act insofar as that act could be traced back to the motives he had identified. He formulated abstract laws that were just as certain as those of mechanics. In reality, such laws were vulnerable to all kinds of disturbing factors, in much the same way as Paracetamol is effective against headaches but not if you get up very early after a long night's heavy drinking – to adopt a usefully vivid example from philosopher Nancy Cartwright. Because the laws of political economy could be observed in their pure form only in exceptional cases, despite always being at work behind the fuzzy world of social complexities, Mill described them, just as Whately and Senior had done, as 'tendencies'; political economy could lay claim to firm but not precise laws.

2.11 A priori and a posteriori

Mill distinguished between his introspective method and induction by calling the first the a priori method and the second the a posteriori method. The latter consisted of the gathering of facts, by statisticians for example, but no general principles would flow from them. As we have seen, the a priori method first sought a general principle governing action and then looked at what would actually be done in a concrete situation based purely on that principle. It was then the politician's job to weigh up the various concrete circumstances and to attach political conclusions to the political economist's 'tendency laws'.

In fact, Mill's essay on the definition and method of political economy suggests that it was possible to look at social reality from an armchair. Its truths could be adjusted but not fundamentally affected by statistical data. So Mill was combining a specific definition of the economically active human being (*homo economicus*) with a specific research method, thereby shying away from the inductive scientific approach. He stressed that it would of course be 'absurd' to assume that the limited, abstract human the economist studied was the same as an actual living person, but the abstraction was needed in order for political economy to make general statements. In short, the approach of political economy was not inductive but deductive.

Mill appeared to have achieved the impossible. In his essay, he had shown that the certainty of the principles of political economy was based on introspection. Those principles allowed the political economist to derive economic laws that were certain, even though they could not be reliably observed with the aid of statistics; economics was therefore a science of tendency laws. So what Whewell abhorred about Ricardo's political economy, its deductive and abstract nature, was actually a virtue, a product of the special character of the 'moral sciences'. Ricardo's political economy was just as scientific as the natural sciences. Mill had transferred the basis of Ricardian political economy from the specific character of land as a production factor to the special character of human dealings. With Mill's essay, political economy actually became a social science for the first time, centred upon the consequences of human actions in concrete economic situations. At precisely the moment when support for Ricardo's political economy was ebbing, Mill gave it a new basis, which was further reinforced by Mill's own *Principles of Political Economy* (1848), in which he systematically brought together all the principles of classic political economy, richly illustrated from historical sources.

2.12 In conclusion

In this chapter, we have seen that controversies about the correct method of practising economics in the nineteenth century cannot be separated from the political controversies of the time. Attempts to neutralize those controversies turned on the question of whether political economy was an inductive or a deductive science, in other words whether it was based on an extensive, statistical description of reality that was used in an attempt to find general laws, or, by contrast, started from first principles, which it deployed to derive conclusions deductively and with unshakeable certainty. This was directly connected to the question of whether political economy was a positive or a normative science: did economists study existing reality or did they prescribe how reality ought to be? These questions were not of a purely academic nature but connected with the most urgent issues of the time: how to think about the relationship between church, state and science; what to do about population growth; what to do about the Poor Laws; what to think about the relationship between different social classes.

Mill's ideas about the scope and method of political economy took the discussion of these issues to a new level. Political economy was a science that dealt with one limited aspect of human dealings. The validity of the laws it formulated was certain (because of introspection), but in concrete cases disturbing factors meant they could not be observed. It remained the job of politicians to weigh up, in practice, those many 'disturbing causes', as Mill called them, based on experience. In the nineteenth century, introspection and complexity became the two concepts used to indicate the specific and unique character of political economy. Mill managed to give theoretical legitimacy to the scientific status of political economy and, for most of the rest of the century, its practitioners endorsed his view, verbally at least. In Chapter 3, we will see, however, that this theoretical basis was not necessarily in keeping with what economists did in practice, even in the case of Mill himself.

References

Hausman, Daniel (1992). *The Inexact and Separate Science of Economics*. (Cambridge: Cambridge University Press).

Maas, Harro (2005). *William Stanley Jevons and the Making of Modern Economics*. (Cambridge: Cambridge University Press).

Meek, Ronald, L. (1950). The Decline of Ricardian Economics in England. *Economica*, 17(65), pp. 43–62, 56.

Wordsworth, William (1970 [1805]). *The Prelude; or, Growth of a Poet's Mind*. (Oxford: Oxford University Press), Book XII, lines 74–80.

3 Economics and statistics

3.1 Introduction

Mill's vision of the method and content of political economy determined the discipline's methodological framework for much of the nineteenth century. This changed in the early 1870s with the rise in Britain, France and Austria of what became known as utility theory. This new theory linked the value of a product not to the value of labour, or to the number of man-hours needed to produce it, but instead to the utility attributed, subjectively, to that product. This notion still lies at the root of the idea current in modern micro-economic theory that individuals act in such a way as to maximize their utility (or preferences).

As well as representing a fundamental theoretical shift, the new theory brought about a transformation in the practical methodology of economics as a discipline. It happened at a time when economists in the German-speaking countries were setting off in a different direction from their colleagues in Britain and France, who stressed that – in contrast to Mill's view – there was no real difference between the ways the natural and the social sciences should be practised. According to British and French economists, Mill had been right to emphasize the complexity of the social sciences, but he had failed to take proper account of the fact that this same complexity was present in the natural sciences, including precisely those disciplines that sought regularity in large statistical data sets, such as astronomy or meteorology. It is no accident that modern probability theory has its origins in astronomy. One of the founders of modern statistics, Adolphe Quételet, whom we met in Chapter 2, used the normal distribution, a concept derived from astronomy, to construct a notion of the 'average man', a standard reference point in his approach to social statistics. Quételet called his approach to social reality *physique sociale* or 'social physics'. Following in his footsteps, late-nineteenth-century economists, especially in Britain, ceased placing their faith in introspection as Mill had done and instead relied on the methods of the natural sciences: mathematics, statistics and experimentation.

The British founder of marginal utility theory, William Stanley Jevons (to whom we shall return later in this chapter), was scornful of Mill's attempts to defend political economy as a science that could be practised without the equipment available to the experimental scientist. He wrote:

> What will our physicists say to a *strictly physical science*, which can be experimented on in the private laboratory of the philosopher's mind? What a convenient science! What a saving of expense as regard of apparatus, and materials, and specimens.[1]
>
> (Jevons, 1971, p. 215)

Francis Ysidro Edgeworth, without doubt the most important mathematical innovator in nineteenth-century economics, likewise ridiculed John Stuart Mill's faith in 'introspective marks of brain-activity'. Edgeworth believed that Mill would do better to turn to the experimental methods of physiologists and psychologists, including those used by Germans such as Hermann Helmholtz, Ludwig Fechner and Wilhelm Wundt, who were transforming psychology into a wholly new, experimental science.

But in the German-language area itself, the debate about the methods to be used by economists was moving in the opposite direction. Until the late eighteenth century, political economy in Germany had been dominated by *Kameralwissenschaft*, with its statistical approach. In the nineteenth century, the Historical School, the successor to this 'cameralism', produced comparative statistical studies intended as material in which the rules governing society could be discerned by induction. The Historical School stressed the culture-specific and therefore national character of such rules.

The founder of marginal utility theory in Austria, Carl Menger, voiced strong and direct objections to the approach of the Historical School. He stressed that economics used a quite different method from that of the natural sciences, precisely because it was based on specific and subjective evaluations by individuals. This meant that economics must not take comparative statistical studies as its starting point but go to work a priori as John Stuart Mill had done. The Austrians stressed more emphatically than Mill that economics was a deductive science that took the logical structure of human dealings as its starting point. The first principles of economics were not psychological in nature but logical. Whereas the marginalists in Britain (and to a lesser degree France) placed the emphasis on empirical statistical research, the Austrians dismissed such empiricism almost in its entirety. In 1932, following the Austrian lead, British economist Lionel Robbins formulated his famous definition of economics as 'the science which studies human behaviour as a relationship between ends and scarce means which have alternative uses' (Robbins, 1981). Robbins did not reject empirical research altogether, but he clearly regarded it as of secondary importance, as merely a means of illustrating 'high theory'.

As a consequence, this chapter falls into two parts. In the first part, I draw upon the work of two nineteenth-century British economists, John Elliot Cairnes and William Stanley Jevons, to show how economics moved from a historical or discursive method to a mathematical, statistical approach. The second part of the chapter shows how the Austrian School attempted to place empirical research outside the 'pure' science of economics. Lionel Robbins' *An Essay on the Nature and Significance of Economic Science* (1932) consolidated the analytical method in economics at the expense of the empirical statistical approach.

The divide between inductive and deductive economics seen in the nineteenth century therefore returned in the 1930s, but in a very different form. The inductive method was by then strongly associated with the deployment of mathematics and statistics, whereas the deductive method involved analysing concepts from first principles. Each produced a very different image of the economist. Was he the type of scientist who derived his knowledge and status mainly from an impressive rhetorical talent and a thorough knowledge of texts? Or was he the type of scientist who distinguished himself primarily by his command of a technical, mathematical apparatus?

3.2 Transformation of the method of economics in Victorian Britain

Around 1863, Irish political economist John Elliot Cairnes wrote a letter to English political economist William Stanley Jevons saying that he had been pleasantly surprised to discover that in researching the consequences for the price of gold of the new finds in Australia and California, Jevons had reached the same conclusions as he had himself, despite the fact that Jevons applied 'entirely distinct methods of inquiry'. Cairnes (1823–74) was twelve years older than Jevons (1835–82) and already a leading political economist, having come to prominence as a result of his highly regarded book on the origins of the American Civil War, *The Slave Power* (1862). In 1863, Jevons was still a student of political economy at University College London, an institution that Jeremy Bentham and James Mill had helped to found.

In the first part of this chapter, Cairnes' work will help to demonstrate how political economists approached their subject before statistical data was presumed to take the form of the quantitative statistical material used by economists today. When empiricism became identified with statistics (in the work of Jevons, for example), the profile of the 'good' economist changed. Like John Stuart Mill, Cairnes was schooled in what we now call the arts or the humanities; a thorough knowledge of history and the classics was regarded as a precondition of being a good economist. We see this reflected in the way the economists of the time set about their work.

Jevons' background was very different. In the years 1850–3, he studied mathematics and 'experimental philosophy' (chemistry) at University College London. He then spent several years working, for an excellent salary, as chief assayer at the Sydney Mint. He was in Australia during the period of the gold rush and, after five years, he returned to London to continue his studies, no longer in the experimental natural sciences but instead

Economics and statistics 23

in political economy. By the time he completed his degree in 1863, he was recognized as a promising thinker. Clearly less at ease in the humanities, the young Jevons published work on instruments such as the barometer and the scale balance, and on logic, meteorology and social statistics. The calculating machines designed by that contrarian of British science Charles Babbage (which are generally regarded as mechanical precursors to the modern computer) inspired Jevons to construct a logic machine, resembling a piano, that could draw logical conclusions from a limited number of statements.

To political economists such as Mill and Cairnes, such subjects lay outside the terrain of political economy. They did ask themselves how their discipline distinguished itself from history and how they could arrive at systematic insights that were on a par with those of the natural sciences, but they did not believe that political economists could use laboratory instruments such as the scale balance, or that the workings of a logic machine would throw any light on the problems of induction that they faced, let alone that a political economist needed to have any knowledge of how such instruments worked. Political economy was a quite different kind of science. The type of mathematical statistics applied to society by Quételet did not enter the picture for a moment.

3.2.1 Cairnes' notes on the condition of the Irish economy

In gauging the difference in method between Cairnes and Jevons, it is useful to look at two pages of one of Cairnes' notebooks. Figure 3.1 shows the notes kept by Cairnes in the mid-1860s on the socio-economic situation in Ireland. He made his notes at the request of John Stuart Mill, who used them for the sixth edition of his *Principles of Political Economy*. Cairnes conveyed his observations to Mill not in the note form shown in Figure 3.1 but as a properly constructed essay based on them. He used the same material in a series of articles he wrote for *The Economist*.

In making his notes, Cairnes followed traditional practice, which involved placing initial observations on the right-hand page. Some are based on primary sources, such as the periodicals that published statistical or other scientific data, while others reflect material found in newspapers or elsewhere. Cairnes 'numbered' his notes alphabetically, a standard practice that had been explicitly recommended by seventeenth-century philosopher John Locke. Note c, for example, is an extract from an article by Irish political economist Neilson Hancock, published in the magazine of the Dublin Statistical Society, about the area of land under cultivation in Ireland in 1860 as compared to the year of the great famine, 1847. Cairnes adds a brief calculation concerning the increase in the amount of land under cultivation and the accompanying growth in the harvest in the same period. So Cairnes did use statistical information, but always in combination with other forms of expression. Statistics could not stand alone. Cairnes did not think in numbers.

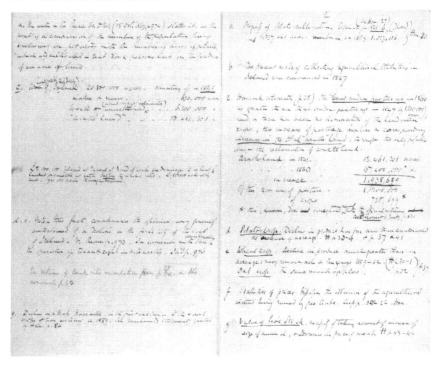

Figure 3.1 Two pages from John Elliot Cairnes' notebook about the state of the Irish economy.

Source: Cairnes papers, MS 8983: 'Economic Notes on Ireland made by John E. Cairnes for John Stuart Mill' (1864). Courtesy of the National Library of Ireland, Dublin.

On the left-hand page he supplemented his primary notes with commentary of his own, or with additional notes from other sources, as in the case of 'potato' and 'wheat crop'. One of his additional notes, for example, is from *Revue des Deux Mondes*, which was among his favourite magazines, and he adds critical remarks of his own about the prevailing public opinion on the subject of the connection between the Irish tenancy system and agricultural productivity. This was the method Cairnes used to develop his notes into the fluent, cohesive text that he sent to John Stuart Mill and then used for his articles in *The Economist* of 1865.

Cairnes could be said to have entered into a dialogue with his sources, with the work of other political economists (he mentions, for example, a speech by Mountiford Longfield) and with the general public, before arriving at a judgement about the development of agriculture in Ireland. He gradually synthesizes diverse material into a coherent article, connecting the potato harvest, the fertility of the soil and the system of tenant rights in Ireland. He does not restrict himself to Irish sources but makes repeated use of the French magazine *Revue des Deux Mondes*, which was much quoted at the time. This kind of dialogue presumes a certain readership; Cairnes was not producing merely a statement of facts but an argument as part of a public debate.

We can see from Cairnes' way of making notes that although he was not averse to using quantitative statistical data, he did so in conjunction with sources of other kinds in order to construct an argument. He was seeking a coherent interpretation of events, and in doing so he combined institutional and quantitative data with the supposed behavioural characteristics of the Irish. This kind of coherent interpretation was a 'tendency' in the sense in which John Stuart Mill used the term.

In the final analysis, Cairnes' way of working does not differ greatly from what William Whewell thought of as an 'inductive' method of practising economics. In the few reflections on the subject in Whewell's notes, he writes of 'history, tracts and conversation' as the material on which the economist based his inductions – precisely the kind of thing Cairnes used in making his notes. In his remarks on the method of the political economist, Cairnes shows himself to be a fervent adherent of Mill's a priori method but, in practice, there was no great difference between what Whewell had in mind and what political economists such as Cairnes, or indeed Mill, actually did. It is therefore unsurprising that Whewell wrote of Mill's *Principles of Political Economy* (1848) that he could not really find much to criticize in it.

3.2.2 The political economist as detective and playwright

So in the nineteenth century, there was not such a clear-cut difference in practice between induction and deduction in Britain as Chapter 2 might lead us to believe. Mill had good reasons for stressing that the a priori method was a combination of induction and deduction. What Cairnes wrote in his notebooks resembled in written form the monthly discussions he led after dinner at, for example, the Political Economy Club. A political economist presented his empirical material as a playwright might, by entering into a dialogue with his sources and interlocutors. Ultimately he was like an omniscient narrator, shaping the material available into an irresistible argument.

The result was an intervention in the public debate. Articles published by political economists such as Mill or Cairnes in one of the many magazines available to them served to mobilize public opinion on important subjects or, even better, to influence politicians in the House of Commons. Longer works appeared in book form. One good example is Cairnes' study of the slave economy of the southern states of America, which he published in 1862 on the recommendation of John Stuart Mill as *The Slave Power*. With that book, Cairnes managed to convince the British government and the British public that the American Civil War was not really about the southern states' right to self-determination, or their putative defence of free trade in the face of the protectionist north – the prevailing view in the British press – but about the preservation of the slave economy. Later, historians expressed all kinds of criticisms of Cairnes' book, but they had few quarrels with his views on the central matter of slavery. Although it was not a major issue in the British public debate about the American Civil War, Cairnes' book successfully made clear to his Victorian contemporaries that slavery was 'the elephant in the room'.

As in his shorter works, Cairnes used a combination of newspaper articles, statistics, exchanges of letters and the US Congressional Record to shape the plot of his book. He believed that the strength of his argument ultimately depended on offering convincing insights into the motives of the people who were active in his economic drama. His cast of characters consisted of slave owners, slave traders, the impoverished white population and the slaves themselves. *The Slave Power* laid bare the structure of the socio-economic drama that inevitably brought the southern elite into political and military conflict with the northern states, where slavery was banned. Cairnes' book is an example of the best a political economist could offer the public. John Stuart Mill wrote an approving foreword, which helped to bring Cairnes useful support in his ambition to become professor of political economy at University College London.

In the book, Cairnes compares his way of working with that of the famous French palaeontologist Georges Cuvier, who was able to use his broad knowledge of palaeontology to identify an animal based on a single bone. In the same way, a political economist could bring his wide reading of economic tracts and pamphlets to bear on the available evidence and thereby compose a coherent picture of the economic phenomenon he was studying. In the case of the southern states, the 'bones' Cairnes found (such as the kinds of crops grown in the region, or the fact that slave owners preferred to keep slaves ignorant rather than to educate them) were used by him as 'clues' from which he could infer 'almost a priori' the character of the southern slave economy.

Cairnes was therefore enthusiastic about the work on induction by Cambridge astronomer John Herschel, who wrote that a 'perfect observer' must have extensive knowledge of his own field and of other neighbouring fields as well. Only then would he be in a position to notice the exceptions that were the key to discovering new laws and connections.

Anyone reading such descriptions of how a scientist sets to work might well find himself thinking of Sherlock Holmes who, in his Baker Street apartment, carefully dosing himself with cocaine or morphine, could solve crimes on the basis of a single clue. Conan Doyle has Holmes speak of his 'little deductions', which enable him to solve the most unfathomable murder cases. Holmes focuses on finding a coherent interpretation of events. To a political economist such as Cairnes, such an interpretation was ultimately all about people's motives for acting, and to him this meant above all how 'self-interest' – an axiom as incontestable to the political economist as gravity was in mechanics – operated in practice.

3.2.3 A graph by William Stanley Jevons

Let us now take my second example, one of the many loose sheets of paper on which, using numbers, William Stanley Jevons gives a series of coordinates (see Figure 3.2). He draws dots between two axes, assigns a number to each dot and then puts a dashed line 'through' them. It is unclear what the dots, or indeed the numbers on the axes, relate to. Perhaps we are looking at astronomical observations of the orbit of one of the planets, or perhaps measurements made in the laboratory, showing a reduction in temperature or changes to the pH level of a fluid. They might equally well concern the movement of a pendulum.

Whatever it represents, the graph suggests measurements in one of the fields of natural science. The dashed line demonstrates Jevons' attitude to his data; he does not make it pass through all the dots but instead draws a flowing line that treats the dots as observations with a certain margin of error. In other words, there is scope for slight aberrations. Jevons thought about his data with the mind of an astronomer and statistician. He draws a line that evens out errors, as an astronomer might. The dashed line 'fits' the data. If we compare this graph with Cairnes' notes, it becomes clear that Jevons' way of thinking and acting is far removed from the practices of political economists such as John Stuart Mill or John Elliot Cairnes.

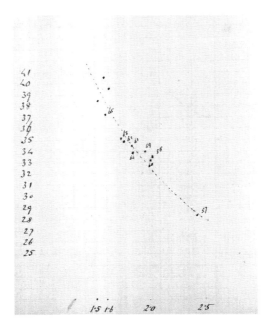

Figure 3.2 A graph by William Stanley Jevons, probably of prices (horizontal axis) and quantities (vertical axis).

Source: Jevons archive item JA/48/89. Courtesy of the John Rylands University Library, Manchester.

The graph probably represents an attempt by Jevons to find a functional connection between the price of a product and demand for it. In contrast to our modern convention, the price is on the horizontal axis and the quantity on the vertical axis. The numbers next to the dots indicate the years for which Jevons had data. To make a graph like this, you would need to be proficient at gathering information about prices and know how to deal with gaps in the data, how to find averages for separate pieces of price data, how to create scales on the two axes, how to plot data in the space between the axes and how to read a functional connection from the dots thus plotted.

In short, making a graph like this demands quite different skills from those used by Cairnes in compiling his notes about Ireland or in structuring his book on the American slave economy. In an essay written to mark the hundredth anniversary of Jevons' birth, John Maynard Keynes wrote in 1935 that Jevons was probably the first economist to examine statistical data with the eye of a physicist or astronomer. In that sense he was a polymath. As much at home in economic theory as in the 'black arts of inductive statistics', Jevons looked at his material with the 'prying eyes and fertile controlled imagination of the natural scientist' (Keynes, 1936). Jevons' background in the experimental sciences and his interest in socio-economic issues led him to deploy his familiarity with scientific instruments and skills in a field from which they had traditionally been seen as remote: political economy.

Cairnes showed great skill at hunting out literary and statistical sources on the subject of Ireland or America. In the case of Ireland, he combined his sources to demonstrate the influence of institutional factors on the population and on production. Jevons' way of working was utterly foreign to Cairnes, and to Mill. Whewell's *Philosophy of the Inductive Sciences* (1847) had suggested that a graph might be an appropriate instrument for use by a political economist in discovering regularities in statistical time series, but he did not develop that idea further. In 1871, two years before his death, Mill wrote in a letter to Cairnes that Jevons was not untalented, but that he had 'a mania for encumbering questions with useless complications, and with a notation implying the existence of greater precision in the data than the questions admit of' (Mill, 1871). Jevons' use of statistical data shows that the political economist was entering new and uncharted territory. Having been a conversation partner in debates, public or otherwise, he now became a specialist in the development of statistical instruments.

3.2.4 The economist as instrument maker

Cairnes realized the full extent of the difference between their approaches when he read Jevons' 1863 study on changes in the price of gold. Discoveries of gold deposits in California and Australia led Cairnes to investigate the effects of the increased supply of gold on prices and production. Building on quantity theories of money developed by David Hume and Richard Cantillon, he argued that on its journey to Europe the new gold would cause prices to rise. His analysis was based on the responses of buyers and sellers of goods to the increased amount of money in circulation.

Jevons ignored such responses completely. Instead of attempting to achieve ultimate insight into how the price of gold would change by looking at the motives for human action, the institutional characteristics of the gold market and so forth, he collected a statistical dataset for the price of some forty different products over a period of twenty years. In his day, the value of the currency was expressed in terms of a fixed gold value, which meant that every price was implicitly related to the price of gold. Jevons' reasoning was as follows: if prices rise on average, then the price of that to which they are implicitly compared – gold – has fallen in relative terms. That is all. We need say no more than that.

Nowadays, we describe a rise in average prices as an increase in the price index, or as inflation, a phenomenon that seems as real to us as protons, neurons or lymphocytes. Jevons was one of the first to construct such a price index. He was well aware that a price index is a different kind of thing from a pile of sandwiches but, even though you could not see, feel, smell or taste a price index, it was a useful instrument in that it could give quantitative precision to a line of reasoning.

What after all would be more probable? That changes in the price of a variety of products as the result of a variety of factors would all end up being in the same direction? Or that price changes in that same variety of products were all in the same direction as the result of a single cause: a fall in the price of gold? The latter, of course. 'The average must, in all reasonable probability, represent some single influence acting on all commodities' (Jevons, 1884, p. 122). The ways in which economic phenomena were changing could be read with more precision from a rise or fall in the price index than from a subtle, qualitative description such as Cairnes offered his readers. Whether or not an index figure pointed to something 'real' was irrelevant. An instrument could take the place of a complex argument about the motives of buyers and sellers in the market.

With Jevons' price index, it was no longer necessary to look at the specific characteristics of different markets. The fact that individuals reacted differently because a market was arranged differently in an institutional sense was no longer an important factor. Up until this point, political economists had paid attention to precisely such factors in their efforts to achieve a coherent interpretation of events, and it was therefore with reference to such factors that they criticized Jevons' study. Cairnes' colleague and fellow Irishman Cliffe Leslie stressed that political economists should study a 'plurality of causes' rather than one simple cause. He meant that the data available to economists were diverse in nature and origin, much like historical data.

Unlike Leslie, Cairnes recognized that in this specific case such criticism missed the central point of Jevons' new approach. Jevons' price index showed that information about individual markets, price developments and institutional and legal orders was not necessary if you wanted to say something about changes in the price of gold. But Cairnes believed that as soon as Jevons applied this same kind of reasoning to the acts of human beings, he was making the mistake of dismissing far too easily the complexity of social reality, the diversity of human motives and the problem of free will.

Jevons responded to this criticism as a natural scientist, saying that such problems presented the economist with the same difficulties as were faced by the astronomer or chemist: every observation involved a margin of error. Free will was evened out for the 'average man'; the law of large numbers renders up a regularity that might not exist for individuals; irregularity at an individual level should not prevent the political economist from looking for regularity in large datasets.

So Jevons transplanted the way in which economists such as Cairnes were used to arguing from a terrain on which the approach of the historian was appropriate to a very different terrain, where quite different skills were required. To Jevons, the 'empiricism' according to which economists developed theories and tested them against quantitative, statistical data no longer dealt with the diverse sources used by historians. Economic theories could be expressed in functional connections – in mathematical relationships – that could be distilled from such data. The political economist changed from being someone who drew his expertise from a profound familiarity with the specific institutional and historical peculiarities of a given economy into someone who needed to become an expert in techniques for uncovering secrets hidden in statistics. Induction was no longer about bringing together information from totally different sources; it was now about handling statistics. The economist was transformed from a detective and playwright into an instrument maker.

3.3 The statistician versus the analytical economist

In the Netherlands, the national economists' association is called the Koninklijke Vereniging voor Staathuishoudkunde. Oddly, its title does not contain the word 'economics' (*economie*) but instead *staathuishoudkunde*, which might be translated as 'state housekeeping'. There are historical reasons for this, and they point to a difference in background between British political economy and that of continental Europe, especially its German-speaking region. In Britain, political economy developed hand in hand with thinking about the place of the free market in society, and with liberalism more generally. On the continent, by contrast, and especially in Germany, there was a tradition of statisticians being commissioned by the principalities of the time to construct extremely detailed socio-economic studies. As we have seen, the word 'statistics' contains the word 'state' (Latin: *status*), which indicates both the etymology of the word and the fact that such studies were used to describe states. A prince could use them to perfect his domestic and foreign policies as far as possible and to maximize his control over his own people and resources.

The Dutch term *staathuishoudkunde* has exactly these connotations, and even long after the Second World War there was still talk of professors of 'state housekeeping' rather than economics. The studies carried out by statisticians involved tables showing population, income and so forth, and there was even a connection with the development of the calculus of probability. To take one example, in the seventeenth-century Dutch Republic, Christiaan Huygens was among the first to calculate the chances of dying based on mortality rates. Grand Pensionary Johan de Witt was another prominent figure who worked on calculations of this kind. Statistics were a matter of national importance.

Let me describe two of the more obvious aims of such statistical studies. First of all, they served to render up useful insights into the total productivity and population of a principality, which could then be used, for instance, to determine the tax base. Second, a detailed picture of the prince's state offered insight into his power, not least by indicating the size of army he could hope to raise. A diagram from the work of the German scholar August Friedrich Wilhelm Crome (1753–1833), professor of *Staatswissenschaften und Kameralismus* at the University of Giessen in the state of Hessen-Darmstadt, arises out of a combination of these two aims. Figure 3.3 shows the absolute and relative size of the populations and state incomes of various European countries. The data used to produce such diagrams might sometimes be extremely politically sensitive, so it was not unusual for a prince to allow statistics to be compiled only under certain strict conditions, or only in secret. Deliberate distortions were quite common too, as a means of fooling the enemy or misleading spies and diplomats.

The research produced by German 'state' or 'cameralist' experts typically took the form of comparative studies of countries, serving to indicate the political latitude available to the state. This made it very different from research by political economists in Britain, who ever since Adam Smith had identified the defence of free commerce as their main goal. We have already seen that British political economists were averse to the 'political arithmeticians', as they called the statisticians of the German-speaking world. Indeed, they felt a general distaste for the tendency of statisticians to put the interests of the state above the blessings of free trade. There was a close connection between continental economics and nationalism and protectionism, as can be seen, for example, in the work of nineteenth-century political economist Friedrich List. It was one of the reasons why John Stuart Mill regarded statistics with disfavour.

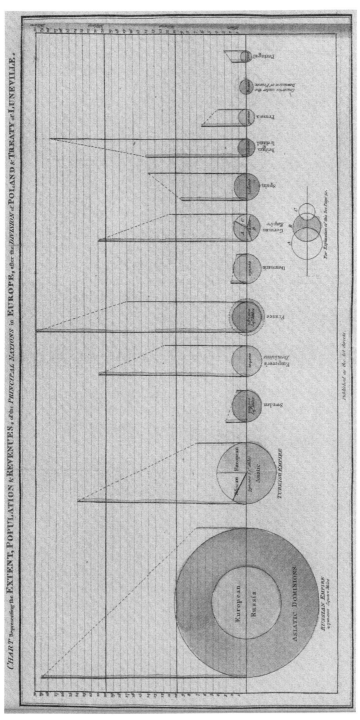

Figure 3.3 A diagram that shows the surface area, population and tax proceeds of the major European states in 1801. Adapted from a diagram by German cameralist statistician Crome by the English pioneer of diagrammatic representation William Playfair, this diagram is misleading in that the gradient of the dotted line appears wrongly to suggest a connection between population and tax take.

Source: William Playfair (2005). *The Commercial and Political Atlas and Statistical Breviary*. (Cambridge: Cambridge University Press), pp. 48–9. The book was originally published in 1801.

In nineteenth-century Germany, this typically continental form of economics was referred to as the Historical School. Its advocates set their faces against the British approach, which they disparaged as *Manchesterthum*, a label derived from the city where the worst excesses of capitalism were becoming evident. For all its limitations, Friedrich Engels' description of Manchester in his *Condition of the Working Class in England* of 1845 is still one of the most important sources for anyone attempting to understand the continental perception of England at the time of its rapid industrialization. In the same period, French political thinker and historian Alexis de Tocqueville wrote that in Manchester 'civilisation works its miracles, and civilised man is turned back almost into a savage' (Tocqueville, 1958, pp. 107–8). In 1843, the average life expectancy of a working-class Mancunian was seventeen years, compared to thirty-eight for a man of the middle class.

One defining feature of the Historical School was that a statistical and a historical approach went together. Statistical studies served to indicate the differences between developments in different countries and to stress the 'national character' of such differences. The emphasis was placed on the excesses of the market and the advantages of state intervention over free trade. Important representatives of the Historical School were Wilhelm Roscher, Karl Knies and Werner Sombart. Max Weber, one of the founders of sociology, was a pupil of Roscher, whom he succeeded as professor of economics at Heidelberg in the late nineteenth century. But as time went on, Weber developed a greater appreciation of the work of the Austrian School, which we shall look at next, and, after 1900, the Historical School found itself very much on the defensive. In Germany, statistics and history had gone together for many years, but now statistical–empirical studies in the German-speaking countries faced considerable opposition.

3.3.1 The Austrian School: economics as a priori science

The decline of the Historical School came as a result of the rise of a new, strongly deductive–analytical approach to economics known as the Austrian School. It is associated, above all, with Carl Menger, and was subsequently represented by Ludwig von Mises and Friedrich Hayek. Max Weber and the later Harvard economist Joseph Schumpeter were warm advocates of the ideas of the Austrians.

Along with William Stanley Jevons and Frenchman Léon Walras, Menger is regarded as the founder of marginal utility theory. All three published their most important books around 1870, but whereas Jevons' method, as we have seen, dissolved the boundaries between economics and the natural sciences, Menger put precisely the opposite case. He believed that political economy was not a natural science, partly because of the complexity of its subject but more importantly because human dealings had a quite different structure from the type of objects and events the natural scientist studied. Central to the thinking of the Austrians was the idea that economists studied a particular type of behaviour known in English as 'economizing behaviour'. Max Weber described it in more general terms as 'means-ends rational action'. It involves purposeful action, or intentionality.

The power of this relatively simple idea was that it enabled liberal, market-oriented economists to separate the subject of economics, at a stroke, from the interests of the state. After all, independently of the form of the state and even of the way society was ordered, people faced a problem: the means of achieving a goal are finite. Whatever the institutional order, people have to choose between scarce means. What could be more obvious than to assume that people forced to choose will try to make the best choice possible? In other words, humans display optimizing behaviour. This reasoning served for some adherents of the Austrian School (such as Ludwig von Mises) as a description of the actual state of affairs. To others (including Max Weber), it was a norm from which people might deviate in their actual behaviour (for example, because they made mistakes, failed to take full account of the situation they were in and so on), although historically, as a result of the rise of capitalism, its validity as a description was increasing.

However that might be, both Weber and Von Mises believed that empirical statistical research was not really the proper way to analyse such means-ends situations. Weber, for instance, had little respect for men such as William Stanley Jevons and Francis Ysidro Edgeworth, who followed the psychologists in relying on experimental methods of studying the ways people make choices. The economists of the Austrian school argued that the economist wishing to study economically relevant behaviour needed no complicated equipment and no laboratory. What they thought of as the incontrovertible facts of everyday experience were sufficient to allow the political economist to draw the conclusion that people strive to optimize different competing needs.

3.3.2 Lionel Robbins: economics as a pure science of concepts

British economist Lionel Robbins (1898–1984) took the same view. Strongly influenced by the work of Max Weber, in his famous *An Essay on the Nature and Significance of Economic Science* (1932) he arrived at a definition of economic science as the study of choice in conditions of scarcity. Like the economists of the Austrian School, he was sceptical about the importance of empirical statistical research for economics as a science.

Robbins saw more clearly than the Austrians that 'economizing behaviour' was actually about two different issues. People did not just have to make a choice between given means to achieve a given end, they also set themselves several goals for which only a limited number of means were available and therefore they had to choose which goal took priority. Robbins believed that the first problem was purely technical, just as the question of what means are to be used to build a bridge is a technical problem. You try to do it with the least possible expenditure of effort and other resources, in other words as efficiently as you can. Only the second problem should concern economists. A problem in an economic sense arose only when a person could not fulfil all his needs with the scarce means available. The study of such a problem was beyond the scope of statistics.

Robbins presented two arguments to support this view. First, preferences are based on individual and subjective evaluations of options, so there is no instrument that can be used to observe the ordering of preferences. Economists such as Francis Ysidro Edgeworth, who believed pleasure and pain were as quantifiable as temperature, were mistaken; there was no device like a thermometer that could measure individual appraisals of value. Second, an appraisal of this kind was dependent on the individual's expectations of the future. Robbins believed it could only be misleading to imagine that we can observe what someone thinks is going to happen. Although perhaps important for economic science, statistics would not help to solve the kinds of questions that interested the economist, which involved intentional acts. As soon as expectations and preferences came into play, statistics became insufficient as a tool.

Robbins illustrated these arguments by using the example of the herring catch in a specific period (he chose the years 1907–8). Imagine that for the period in question the maximum price of herring was below the market price, and statistical research showed that the price elasticity of demand was −1.3 and there was an excess of two million barrels of herring. 'How pleasant it would be to say things like this! How flattering to our usually somewhat damaged self-esteem *vis-à-vis* the natural scientists! How impressive to Big Business! How persuasive to the general public!' (Robbins, 1981, p. 108).

But what use would such research be? How could we know that the price elasticity of demand would remain at −1.3? Surely it might be changed by factors as diverse as religious ideas, fashion, the number of consumers in the market, falls or rises in income and so on. There was no reason at all to assume that a price elasticity of −1.3 was an expression of a constant law of nature. At best, therefore, such empirical studies made a contribution to the writing of economic history about a specific period. They could not be of any use when it came to deriving general conclusions, so they could not help to develop economics as a science. In short, Robbins believed that what was known as the *ceteris paribus* condition ('all other things being equal') had not been fulfilled. The impossibility of fulfilling this condition in practice had been reason enough for Mill to conclude that the experimental method used in the natural sciences was unavailable to political economy. If price elasticity in the past was −1.3, that did not tell us anything about the future. Robbins' verdict on the contribution made by statistical studies to economic science was therefore damning:

There is no need to linger over the futility of these grandiose projects ... They are not new ... It is just about a hundred years ago since Richard Jones ... sounded the note of revolt against the 'formal abstraction' of Ricardian Economics, with arguments which ... are more or less exactly similar to those which have been expressed by the advocates of 'inductive methods' ever since. It is safe to say that, until the close of the War [he is of course referring to the First World War], views of this sort were dominant in German University circles ... Yet not one single 'law' deserving of the name, not one quantitative generalisation of permanent validity has emerged from their efforts.

(Robbins, 1981, pp. 113–14)

I am not concerned here with the question of whether Robbins offers a sound argument against empirical statistical research, nor indeed with the issue of whether or not economics, as a science, should strive to formulate general laws. The point is that Robbins compares all the 'inductive methods' we have examined so far to the 'formal abstractions' of the Ricardian school and then treats the Ricardian method as equivalent to the deductive, a priori approach of the Austrian School. We have seen from the example of John Elliot Cairnes that advocates of Ricardo's theories in the nineteenth century went to work in a way that did not differ greatly from the way historical research had generally been carried out. But in Robbins' hands, political economy was transformed from a discipline that attempted to give a coherent description and explanation of socio-economic reality into a discipline that was more concerned to offer a conceptual analysis of a specific type of behaviour. It therefore shifted from addressing the general public to addressing a purely academic audience.

In Robbins' own work, such analyses were primarily verbal but, even during his lifetime, mathematics became the preferred language for the analysis of behaviour that was aimed at optimization. This use of mathematics was totally different in character from the use to which it was put by Jevons, who deployed statistical data as the basis of a search for an underlying, explicatory functional form. In the twentieth century, such a direct relationship between mathematics and statistics would be sought in econometrics, which was just coming into being in the period when Robbins wrote his book. But alongside it, a whole industry producing theoretical mathematical models has grown up within the science of economics that involves no such close relationship between the two. We shall look at an example in Chapter 7. In the next few chapters, we will concentrate on the rise of econometrics, in other words the building of models in which mathematics and statistics go hand in hand.

As we shall soon see in detail, one consequence of the rise of econometrics was that within economics the concept of empiricism was increasingly identified with statistical data of the kind on which Jevons tested his intellectual capacities. As a result, the field of play on which the economist was active no longer touched upon that of the historian or sociologist, who tries to stage-manage his evidence as far as he can. In the twentieth century, the economist increasingly became a scientist who develops instruments for use in discovering the secrets hidden within statistical data. Economics was becoming mathematical rather than discursive. At the same time, economists were less and less likely to be famous figures who saw it as their task to intervene in the public, political domain. Economists became specialists, like other scientists to be found in academic or quasi-academic institutions, and they now speak primarily to their colleagues. This does not mean that their opinions have disappeared from the public realm. On the contrary, their judgement is still very much present behind the scenes, but it is now veiled by what is presented as objective, impersonal science, with all its many tools and instruments.

Note

1 Unless otherwise stated, in this and other quotations italics are as in the original.

References

Jevons, William Stanley (1971). John Stuart Mill's Philosophy Tested, in William Stanley Jevons, *Pure Logic and Other Minor Works* [1890] (New York: B. Franklin).
Jevons, William Stanley (1884). *Investigations in Currency and Finance*. (London: Macmillan).
Keynes, John Maynard (1936). William Stanley Jevons 1835–1882: A Centenary Allocution on his Life and Work as an Economist and Statistician. *Journal of the Royal Statistical Society*, 99(3), pp. 516–55, 524.
Mill, John Stuart (1972). *Collected Works 17*. (Toronto: University of Toronto Press), pp. 1862–3, letter 1698, 5 December 1871, to Cairnes.
Robbins, Lionel (1981 [1935]). *An Essay on the Nature and Significance of Economic Science*. (London and Basingstoke: Macmillan), p.16 [the first edition was published in 1932, the second edition was significantly rewritten].
Tocqueville, Alexis de (1958). *Journeys to England and Ireland*. (London: Faber and Faber).

4 Business-cycle research
The rise of modelling

4.1 Introduction

With the emergence of econometrics in the 1930s, nineteenth-century research practices lost their scientific patina once and for all. Econometrics, or the building of mathematical models that are then assessed by testing against empirical data, became the standard method of empirical research. Just as astronomers relied on their telescopes, just as a chemist could not do without his balance or a biologist his microscope, so a whole new way of working developed within the science of economics. It became a separate discipline in which instrument-makers developed instruments such as price indexes, graphs and models designed to enable economists to tease out information from statistical data as efficiently and effectively as possible in order to 'measure' economic reality. This new approach required a new kind of scientist and new institutions. In the Netherlands, they were represented by Jan Tinbergen and the Central Planning Bureau (CPB), the government body responsible for economic forecasting and economic policy analysis.

In this chapter, we shall compare the earlier, primarily inductive statistical research into business cycles and market conditions that took place at the Dutch Central Bureau for Statistics (CBS) with Tinbergen's mathematical modelling approach. The CBS concentrated on creating what it called a business-cycle barometer, an instrument designed to show the actual economic climate and to predict how it was going to develop in the future, while Tinbergen's model served as both an explanation of the economic cycle and a forecast of cycles to come. We will return later to the difference between predictions and forecasts.

4.2 Economic barometers in the interwar period

The starting point for this chapter is the business-cycle research carried out by the CBS between the two World Wars. Central to that research was the quest for a statistical 'barometer', an instrument that would indicate current conditions of production and trade the way a barometer indicates air pressure. At the same time, just as in meteorology, such an indicator could also serve to show how economic conditions would develop in the future, in much the same way as a barometer both indicates the current position and predicts future developments.

The idea of developing an economic barometer came from the United States. In the early twentieth century, entrepreneur Roger Babson had developed the Babson barometer, on a commercial basis, as an instrument for anticipating upturns and downturns in the business cycle. Babson sold his barometer by posting a newsletter to interested companies, saying it would enable them to prepare for economic ups and downs. But it was the 'ABC barometer' developed by Harvard statisticians such as W.M. Persons that made the greatest impression in Europe. It was the product of the Harvard Committee for Economic Research, established in 1917, and it consisted of a combination of three graphs, each of them based on several underlying statistical data series (Figure 4.1). The representation of the three curves in a single diagram clearly suggested that the movements of the three separate factors were related. The curves for economic activity and the money market (B and C, the unbroken lines) seem, for example, to follow the curve for the stock market (speculation, the hatched line A), after a slight delay.

Whereas Babson had suggested that his barometer expressed Newtonian, mechanistic laws governing business cycles, the Harvard statisticians stressed that no such laws existed. They saw their barometer purely as an instrument that registered the facts, although it could certainly be of use as a predictive device. Like a barometer in meteorology, the Harvard barometer would both show the current economic situation and give an indication of future developments.

Figure 4.1 The Harvard barometer. This barometer was developed by Warren M. Persons on behalf of the Harvard Committee for Economic Research. The three lines (A, B and C) show, respectively, speculation, business, and money and credit. Each line is itself compiled from underlying time series.

Source: Warren M. Persons (1923). The Revised Index of General Business Conditions. *The Review of Economics and Statistics*, 5(3), pp. 187–95, chart 7, p. 194.

Research into barometers of this type was given a considerable boost by the arrival of John Maynard Keynes as guest editor of the *Reconstruction in Europe* supplement published regularly by the weekly *Manchester Guardian Commercial* in the early 1920s, which reported on reconstruction efforts in Europe after the First World War. At Keynes' request, statistician Arthur Bowley of the London School of Economics convened a congress with participants from various national statistical bureaus to discuss the possibility of their countries jointly setting up an instrument similar to the Harvard barometer. It would be published in the *Manchester Guardian* supplements. He received favourable responses from abroad.

Representing the Dutch, the director of the CBS, Henri Methorst, and the head of its department for business-cycle research, M.J. de Bosch Kemper, took part in these discussions. It was De Bosch Kemper who first set about trying to develop a trade-cycle barometer. The results of research on the matter at the CBS led to an approach to economics that was very different from that of Keynes, as we shall see in Chapter 5. Individual virtuosity in assessing the available sources gave way to systematic gathering of data for the purposes of collective work on a mathematical model.

4.3 De Bosch Kemper's business-cycle research at the CBS

In the early 1920s, the CBS's department for business-cycle research was headed by M.J. de Bosch Kemper, who had trained as an engineer. He belonged to the kind of aristocratic family that, since the nineteenth century, had regarded it as their task to contribute to the edification and improvement of the working class. His grandfather was one of the founders of the Koninklijke Vereniging voor de Staathuishoudkunde (the Dutch equivalent of Britain's Royal Economic Society) and the author of what is now regarded as the first sociology book about the Netherlands. His aunt, Jeltje de Bosch Kemper, laid the groundwork for the domestic science schools that were intended to give an elementary form of education to young Dutch working-class women. De Bosch Kemper occasionally gave talks to unions and to workers' organizations in which he tried to familiarize them with the most important types of social statistics. The education of the working classes was his aim and social statistics the means to achieve it.

At the CBS, De Bosch Kemper was charged with developing a barometer similar to that of the Harvard statisticians. The idea of trade cycles did not come fully into focus until it was expressed in the form of curves in a diagram. A good example of the transition from an approach in which statistical data was presented in tables to one using graphs is the work of French statistician Clément Juglar, whose name is still associated with the fixed investment cycle of up to about twelve years in length. As Mary Morgan explains in her *History of Econometric Ideas* (1990), his 'table graphs' show time on the vertical axis and the changes in balance totals of the French central bank on the horizontal axis. The lines connecting these changes indicate the economy's business cycles.

Until then, socialist Bob van Gelderen had contributed most to the study of business cycles in the Netherlands. In 1913, using the pseudonym J. Feder, he published an article entitled *Springvloed* (*Spring Tide*) in the radical socialist paper *De Nieuwe Tijd* (1913). The title of the article suggests some kind of rule-governed tidal movement in economic life, a notion expressed unambiguously in *Het Economisch Getij* (*The Economic Tide*),

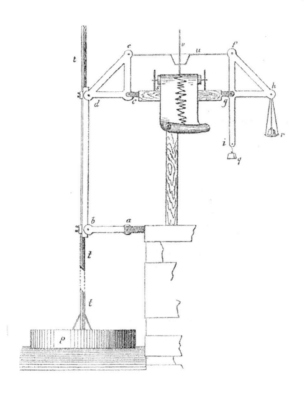

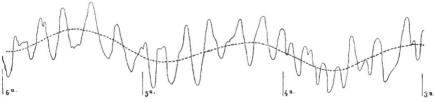

Figure 4.2 Limnimeter and graph from Jules Marey, *La Méthode Graphique*. The limnimeter is a gauge developed by Swiss scientist François-Alphonse Forel to register changes in the surface level of Lake Geneva.

Source: Etienne-Jules Marey (1878). *La Méthode Graphique dans les Sciences Expérimentales et principalement en Physiologie et en Médicine*. (Paris: Masson).

a book about trade cycles published in 1929 by Marxist Sam de Wolff. The tables and graphs used by Van Gelderen in his article were primarily illustrative in character. This would change with the research into business cycles carried out by the CBS. De Bosch Kemper made it his goal to register real economic trends with his barometer, automatically as it were. The idea that graphs could register actual trends was supported by the work of the French professor of physiology Jules Marey who, in his much-discussed 1873 book *La Méthode Graphique* (*The Graphical Method*), advocated the use of graphs as a universal scientific language. One example Marey gave involved something known as a limnimeter, a gauge developed by one François-Alphonse Forel to register changes in the water level in Lake Geneva (see Figure 4.2), 'limnos' being the Greek word for 'lake'. Further examples were the 'indicator diagrams' produced by the inventor of the steam engine, James Watt, who, in the late eighteenth century, added an instrument to the steam engine capable of measuring changes in the engine's efficiency that were otherwise invisible (see Figure 4.3). In the late nineteenth century, strongly influenced by Marey, French statistician Emile Cheysson wrote about the use of graphs, comparing the graphs of international trade with Watt's indicator diagrams. To Marey and Cheysson, graphs were instruments for automatically registering hidden laws and relationships, thereby making them visible.

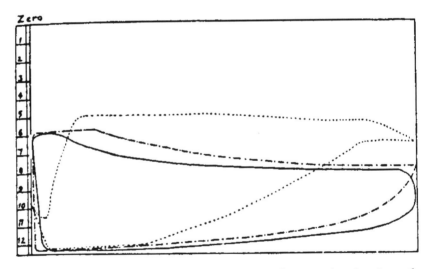

Figure 4.3 James Watt's indicator diagram. The horizontal axis gives the temperature, the vertical axis the pressure. The lines on the graph are the results of different settings of the steam engine.

Source: R.L. Hills and A.J. Pacey (1972). *The Measurement of Power in Early Steam-Driven Textile Mills, Technology and Culture*, 13, p. 41. Portfolio 1381, courtesy of University of Birmingham.

This way of thinking was perfectly suited to the place the CBS ascribed to itself in the Dutch governmental system. A brochure of 1939, aimed at the general public, sketched out the process that led from data-gathering by people external to the organization to statistical representation (see Figures 4.4 and 4.5). Institutionally, the CBS saw itself as a limnimeter, automatically registering invisible economic trends (or trends in society more generally), which its work then made visible to everyone. It was in this spirit that De Bosch Kemper wrote in his retrospective journal *Dagboek conjunctuuronderzoek* (*Diary of Business-Cycle Research*) that in 1921 he had expected his barometer 'would indicate almost automatically in which phase of the cycle [the economy found itself] and what the prospects were for the near future'.

The available statistical data could be represented in the form of business cycles. Its registration would be objective and neutral, not influenced by any political bias or subjectivity in observation. Although in the early twentieth century there was resistance among businesspeople to the centralized collection of data by the CBS, the organization regarded itself as a neutral and objective body, registering developments in economic life. Today, the coding of incoming information is automated; in the interwar years, it still required an entire department manned (or rather womanned) by coders who were expected to be as objective and neutral as the computer is today (Figure 4.6).

For a wide variety of reasons, the indicator dreamed of by De Bosch Kemper was never produced. The Harvard barometer served as an example but, as mentioned earlier, each of the three curves in the diagram was based on several underlying time series. In other words, each curve recapitulated multiple time series. De Bosch Kemper did not succeed in moulding the time series for the Netherlands into a similar convenient threesome. Even showing a clear cyclical development turned out to be harder than expected and required considerable manipulation of the statistical data.

In these manipulations, De Bosch Kemper used purely statistical notions such as 'deviation from the average', in the economically loaded sense of 'disturbances of market equilibrium', but he did not succeed in making a satisfying connection between statistical and economic terminology. So the hope that a barometer could be developed that might also be used to predict the future soon faded. With the arrival of Jan Tinbergen in 1926, research on creating a Dutch barometer was to take a very different turn. Tinbergen drew a clearer distinction between economic and statistical approaches and, at the same time, articulated the relationship between the two far more convincingly. Before we look at Tinbergen's years at the CBS, here is a quick sketch of his background.

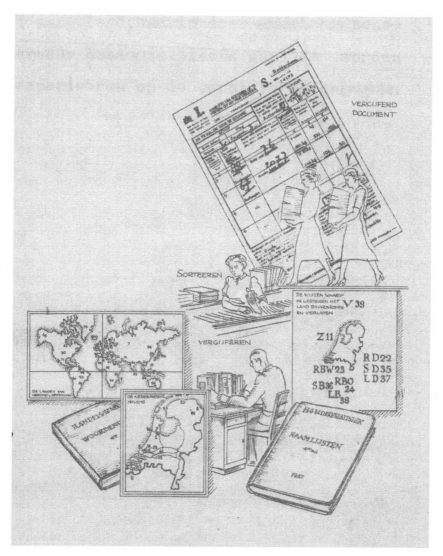

Figure 4.4 Two illustrations from a CBS brochure of 1939 explaining the compilation of import and export statistics. The first illustration shows part of the process of collecting, sorting and transposing the information into figures, the second shows this last procedure in detail.

Source: 'Korte beschrijving van de wijze waarop de statistiek van den In-, uit- en doorvoer wordt samengesteld', CBS Brochure (1939). Courtesy of Beeldbank CBS.

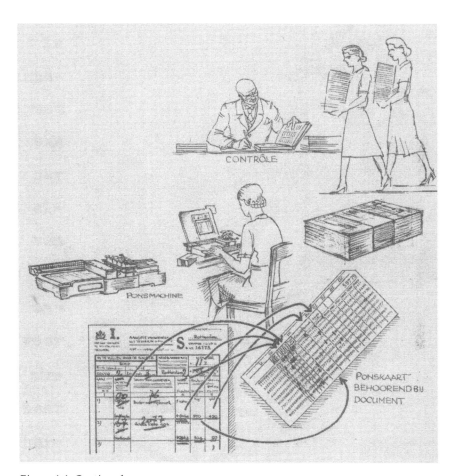

Figure 4.4 Continued

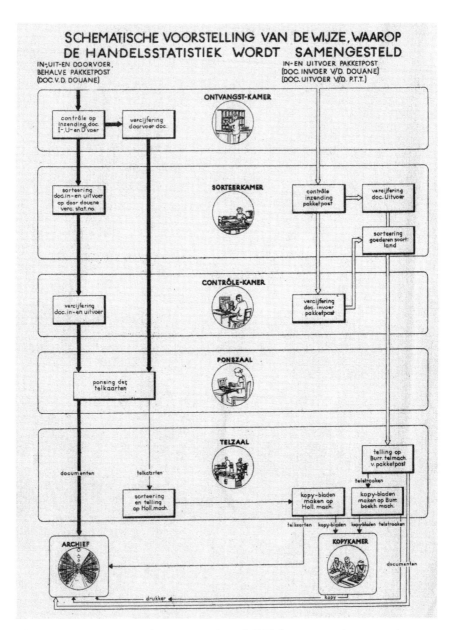

Figure 4.5 Diagrammatical representation of CBS data-processing operations on trade statistics, visualizing the complete internal procedure, from data reception to representing and archiving the original and processed results.

Source: 'Korte beschrijving van de wijze waarop de statistiek van den in-, uit- en doorvoer wordt samengesteld', CBS brochure (1939). Courtesy of Beeldbank CBS.

Figure 4.6 Female employees enter statistical data into the Burroughs computers. The statistical data were on punched cards that coded for different characteristics. Two women worked at each computer, one reading out the quantity of goods per country and feeding it into the machine, which performed the calculation. The other woman did the arithmetic by hand as a way of checking. Every year some two million punched cards were processed. The wires in the photograph are probably electrical cables attached to the ceiling lights.

Source: Ida Stamhuis, Paul Klep and Jacques van Maarseveen (eds) (2008). *The Statistical Mind in Modern Society. The Netherlands 1850–1940*, Part II. (Atlanta: Rodopi), p. 293. Courtesy of Beeldbank CBS.

4.4 Jan Tinbergen: a hybrid expert

In 1926, then aged twenty-three, Jan Tinbergen wrote a letter to F.M. Wibaut, who was both a leading figure in the SDAP (Dutch Social Democratic Workers' Party) and an Amsterdam alderman with responsibility for housing. He asked whether Wibaut knew of 'useful work' that could be done by an 'if necessary extremely sober-living' person 'who has had an economical statistical training':

To be precise I have taken doctoral exams in mathematics and physics, but do not believe I can be of great service to the socialist movement with those subjects. Since my professor is very much in favour of me continuing to study, I am now working at economics and statistics. I arrived at the decision to study those subjects because of the economic character of many socialist reforms and through the conviction that the socialist economies will need skilled people ... With this training, will I be of use to the working class?

(Tinbergen, 1926)

Jan Tinbergen had studied at Leiden University under the mathematician and physicist Paul Ehrenfest, a contemporary and friend of Albert Einstein. Tinbergen was from a highly principled socialist background. He drank no alcohol, refused military service and used what was known as Kollewijn spelling, which meant writing words more or less as they were pronounced so as to eliminate the completely redundant irregularities and elaborations that Roeland Anthonie Kollewijn believed served only to preserve class distinctions.

Having resisted conscription, Tinbergen began his alternative civilian service as a guard at Rotterdam prison, but his father managed to secure a place for him at the CBS, where Tinbergen worked intensively on research into Dutch and international business cycles. During and after the Second World War, he was closely involved with the setting up of the CPB, becoming its first director. In 1969, he and the Norwegian Ragnar Frisch were awarded the first Nobel Prize for Economics for their fundamental contribution to the development of econometrics.

Today's CPB has an official English title: CPB-Netherlands Bureau for Economic Policy Analysis. This gives a better indication of the activities assigned to it, right from its foundation, than the literal translation (Central Planning Bureau). The CPB analyses and predicts economic developments with the purpose of aiding policy preparations. It also analyses the likely economic consequences of intended policies. This conception of its task refers back to the start of Tinbergen's work on economic models. In 1935, he was among those responsible for creating the SDAP-Plan van de Arbeid (the Social Democratic Workers' Party Labour Plan), an ambitious proposal aimed at stimulating the economy through, among other things, large public works projects. In that time of economic crisis, Tinbergen developed a 'model' of the Dutch economy. It was an outcome of his work at the CBS. He investigated various possible scenarios, depending on the policy chosen, and presented them in 1936 in a preliminary report to the Vereeniging voor Staathuishoudkunde en Statistiek (Association for Political Economy and Statistics). He presented his model to the public shortly after the devaluation of the guilder in September 1936, but his model-based analysis confirmed that the abandonment of the gold standard, which the Netherlands was one of the last countries to implement, was a wise policy. We will look specifically at the importance of scenario analysis in Chapter 9.

Tinbergen's letter to Wibaut makes clear that he was a hybrid expert, schooled by one of the most important physicists of his day, Paul Ehrenfest, and that he was determined to use his talents, skills and abilities in the field of statistics and economics, and in the service of the working class. By crossing the boundaries between disciplines, he changed both the image and the operating procedures of the economist.

4.5 Tinbergen at the CBS: the rise of model-building

It was with an almost audible sigh of relief that De Bosch Kemper announced Tinbergen's arrival at the CBS. He was not alone in being impressed by Tinbergen's statistical and mathematical skills, and Tinbergen quickly took the lead in developing the barometer, finding inspiration for his approach in the work of Ernst Wagemann, director of the Institut für Konjunkturforschung (Institute for Business Cycle Research) in Berlin, where he spent several months. Tinbergen abandoned the fruitless efforts of De Bosch Kemper to squeeze Dutch statistics into the straitjacket of the Harvard barometer and instead adopted an approach in which four separate graphs were shown. One important consideration was that unlike the American economy, the Dutch economy was extremely dependent on imports and exports. Each graph included comparative data from the United States, Britain and Germany, so the barometer served above all as a way of looking at the position of the Netherlands in relation to other countries (see Figure 4.7).

De Bosch Kemper had already been forced to abandon his aim of creating a barometer capable of prediction and, with Tinbergen's innovation, the Dutch barometer became purely an instrument for registering data and for giving a comparative assessment of the Dutch economy as it currently stood. This was a long way from the original aims set out for it by De Bosch Kemper, which is not to say that Tinbergen lost sight of the Harvard barometer. Following the example set by Wagemann, Tinbergen interpreted that barometer less as an empirical reflection of the American economy than as a stylized representation of the laws governing economic cycles. Tinbergen began to experiment with what he called 'schemata', 'mathematical machines' or 'models'. By combining a number of equations for, say, supply and demand, he tried to simulate cycles of the kind shown by the Harvard barometer. An article he wrote to mark the tenth anniversary of Berlin's Institute for Business Cycle Research demonstrates the results of this kind of study.

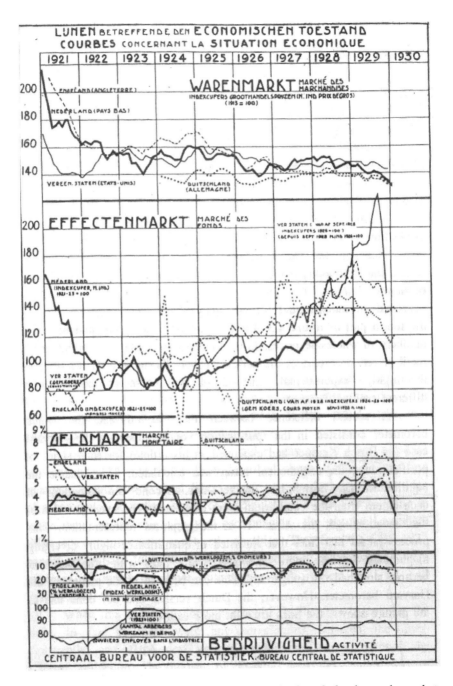

Figure 4.7 CBS barometer comparing fluctuations in the wholesale goods market, stock exchange, money market and (un)employment rate for the Netherlands, Britain, Germany and the United States.

Source: Courtesy of Beeldbank CBS.

Figure 4.8 makes clear that the different economic trends follow each other with a certain amount of delay. This could also be seen from the Harvard barometer but, with Tinbergen's idealized barometers, it was possible to follow precisely the causal connection that produced this effect, with the help of mathematical equations that had the same impacts. Initially intended as a descriptive and predictive instrument, the barometer now became a means of discovering explanations.

Tinbergen's quest for explanatory mechanisms for business cycles comes into even sharper focus when we look at how he experimented with various ways of representing equations that deal with supply and demand. 'Experimented' is my word for it, not Tinbergen's. In his early work, he looked for mechanisms that would best explain the movements of prices and quantities on markets, comparing his diagrams with, for example, statistical data for the market in coffee or potatoes. Tinbergen called these diagrams 'schemata' or 'mechanisms' and only later settled on the term 'model'. As a result of his work on these simple models (and his visit to Berlin), it became clear to him that the development of business cycles depends on the delay in adjustments (or more generally in the parameter values) of supply and demand. Tinbergen's models are therefore mechanisms for generating economic cycles that can be identified in the CBS statistical data.

Tinbergen goes on to illustrate this using a model of Dutch shipbuilding, in which he derives the values of the parameters in the equations (for example of α and β in $y = \alpha x + \beta$) from registered statistical time series. His assumption was that the model would achieve a better representation of 'reality' (which is to say the empirical time series) by being allowed to increase in complexity. The addition of explanatory variables would, he believed, enable the model to become a better 'fit'. As was the case with his simple mechanisms, the addition of variables to a model would have to be based on theoretical considerations.

As his next step, Tinbergen developed a model of the Dutch economy. It was the first ever macro-economic model of a national economy, and the form it took can be seen in Figure 4.9. He captured the Dutch economy in a system of twenty-four 'elementary equations', made up of what we would now call 'behavioural equations' and 'definitional equations'. Some variables are shown with a time lag, which gives dynamism to the system. Tinbergen's model is a complex system of difference equations, all of them based on the time series collected by the CBS and other statistical bureaus. In a way entirely his own, Tinbergen visualized the 'fit' between the model's estimates and empirical time series for the different variables (see Figure 4.10). For each equation, the model calculation, a solid line, is set against the empirical time series for that same variable (the dotted line). In a separate appendix, Tinbergen gave the meaning of the symbols he had used. For example, u stood for 'the total volume of consumer goods produced (or other goods, if for export)'. Equation 11 in Figure 4.9 shows the movement of these variables, and Figure 4.10 clearly shows the difference between model and reality for the variable concerned, while also making clear that the estimate and the actual outcome follow the same course. The model could serve as an explanatory and predictive instrument because it contained the mechanism of the Dutch economy, if in an imperfect form (the estimates and the CBS data did not coincide completely).

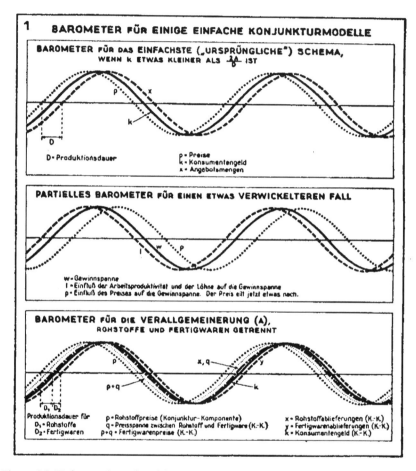

Figure 4.8 Tinbergen's stylized barometer for three different models that move from a simple to a generalized case. The individual graphs refer to different variables in each case, but each separate diagram reproduces the sequential order of the Harvard barometer.

Source: Über den Wert mathematischen Konjunkturtheorien, *Beiträge zur Konjunkturlehre: Festschrift zum zehnjährigen Bestehen des Instituts für Konjunkturforschung.* (Hamburg, 1936), pp. 198–224.

Bijlage C: Overzicht der elementaire vergelijkingen.

1. $\ell - \ell_{-1} = 0.27(p_{-1} - p_{-2}) + 0.15\,a$
2. $p = 0.04\,p'_A + 0.15(\tau'_A + 2\ell - 6t) + 0.08\,u$
3. $q = 0.74\,q'_A + 0.16(s'_A + 2\ell - 6t)$
4. $p_A = 1.23\,p_w + 0.04(\tau'_A + 2\ell - 6t)$

5. $u = u_A + u'$
6. $u_A = z + 2.23(p_w)_{-0.28} - 1.26\,p_A$
7. $u' = L + E' - 2.49\,p$
8. $v'_A + 3y'_A = 0.51\,Z_{-1}$
9. $a = b + 0.20\,u'_A + 0.96\,z'_A$
10. $y'_A = 0.69\,b$
11. $u = 1.72\,u'_A + 4.35\,x'_A$
12. $x'_A - 0.71\,u'_A = -0.42\,p + 0.39\,p'_A$
13. $y'_A - v'_A = 0.86(q'_A - q)$

14. $L = a + \ell$
15. $Z = J + U' + U_A + 3b + 0.71\,q - L - X'_A - U'_A - Y'_A +$
    $+ 0.26\left[s'_A - (s'_A)_{-1}\right] + 0.31\left[\tau'_A - (\tau'_A)_{-1}\right] + 0.47\left[p'_A - (p'_A)_{-1}\right] +$
    $+ 0.3(Z - Z_{-1})$
16. $E = 0.48\,Z + 0.20\,Z_{-1}$
17. $E' + E'_{-1} = 0.26\,E_{-1}$
18. $E'' + E''_{-1} = 1.74\,E_{-1}$
19. $U_A = u_A + 0.68\,p_A$
20. $U' = L + E'$
21. $U'_A = u'_A + 0.58\,p'_A$
22. $V'_A = v'_A + 0.13\,q'_A$
23. $X'_A = x'_A + 0.41\,\tau'_A$
24. $Y'_A = y'_A + 0.13\,s'_A$

Figure 4.9 Tinbergen's model of the Dutch economy, as presented in a memorandum for policy purposes.

Source: 'Nota betreffende het algemeene conjunctuuronderzoek', August 1936, Appendix C. Courtesy of Beeldbank CBS.

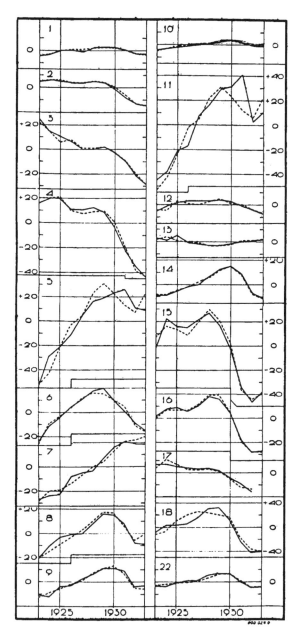

Figure 4.10 Tinbergen's correlation diagram. The dotted lines are registered time series, the solid lines are the model estimates.

Source: 'Nota betreffende het algemeene conjunctuuronderzoek', August 1936, Appendix B, Figure A. Courtesy of Beeldbank CBS.

The importance of Tinbergen's new way of working can hardly be overstated. In ten years, Tinbergen had created something unparalleled in the history of economic science. He had made an empirical model of a real, empirical economy, based on economic theory and on data collected by the CBS and other statistical bureaus. His model brought about a revolution in the way in which economists were able to think about the connection between economic theory and empirical data. Furthermore, it suggested that it was possible to distinguish unambiguously between the two. Empirical data were henceforth associated with statistical datasets, whereas economic theory described the mathematically constant structure – the mechanism – of the economy that could be estimated with the aid of statistical data.

This division of tasks also meant that the work of economists who did not regard 'empirics' as statistical data became increasingly suspect. For the first time, Tinbergen's model-based approach made it a matter of course to see statistical data series as the economist's only valid empirical material. Figure 4.11, which is taken from the more extensive analysis in Marcel Boumans' book *How Economists Model the World into Numbers* (2005), shows the precise division of tasks between the economist and the statistician. The economist reflects upon the mechanism that lies at the root of economic reality, and by doing so renders up the explanatory variables. The statistician supplies empirical statistical data. The work of the econometrician is then to bring mechanism and data together.

In section 4.6, we will see that Tinbergen not only redefined the relationship between theory and data in economic science but transformed the playing field for economic policymaking.

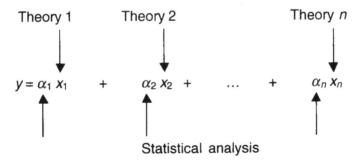

Figure 4.11 The relationship between economic theory and statistics.

Source: Marcel Boumans (2005). *How Economists Model the World into Numbers.* (New York: Routledge), p. 46. Courtesy of Marcel Boumans.

4.6 The model as a political umpire

Tinbergen quickly gained respect in scientific circles, but it is important to remember that he carried out his work in the civil-service environment of the CBS. Although the CBS strove to collect and provide objective data, it was nevertheless an institution that served the public interest. Both aspects, the scientific and the public cause, are clearly expressed in the official documents produced by the CBS's business-cycle research department. To take one example, here is an official document dated 23 October 1935 and addressed to the Dutch minister of economic affairs:

> There is one objection to the business-trend barometers that have been constructed thus far, namely that they do not comprise a closed system of variables, which is to say a system of variables that gives a single, if simplified, picture of reality. In the lecture by Mr. Tinbergen referred to in the memorandum (of 11 June 1935) such a system is announced. From this then follows a programme of statistical research, of which several of the most important points had already been mentioned in the memorandum.

In his 'Memorandum concerning the general research into business cycles' of August 1936, Tinbergen presents for the first time his 'model of Dutch economic life', the goal of which was 'to become familiar with the *causal connection* between the trends of separate economic variables in a *quantitative* sense':

> If one is content for a moment with the fact that the figures produced give the best measurement that can currently be given, then the representations in the form of graphs to be found in appendix B show that the course of an important number of annual indices of Dutch economic trends can be 'explained' with great accuracy through the use of the formulae given in the table in appendix C. These formulae represent in a certain sense (see the reservations contained in what follows) the dynamic laws that govern economic development. The formulae are framed, as far as their general form goes, on the basis of economic observations of a theoretical and practical nature, while the numerical coefficients, insofar as these are not settled a priori, are determined with the help of the (multiple) correlation calculation, at least to the degree this has proved possible.

In a separate section, Tinbergen indicated what the 'significance of the model' was 'for the theory and politics of business cycles'. It rendered up 'guidelines' for the improvement of statistical data collection; it presented the 'possibility of creating more scientifically-based types of barometers' (for example because calculations based on models and the available statistical data diverged); it formed a starting point for 'systematic' research into the 'dynamics of economic life'; and it presented the possibility of making a comparison between these dynamics if the model was 'left to itself', compared to situations in which 'certain economic policy measures are implemented'.

Tinbergen's model was a scientific instrument, a 'barometer'. It could explain events, but it was also a policy instrument. The context in which Tinbergen developed his scientific instrument was the civil service; the CBS was not a scientific research institute, rather it was at the service of the government. By implication, the same applied to Tinbergen's econometric model. Even if economic policy was primarily in the hands of politicians, Tinbergen left no one in any doubt that his model determined the boundary conditions within which it was meaningful to talk about the economic fabric and economic policy. In his memorandum of August 1936 he stressed that:

> For an effective debate it would be of great value if the objections to the model described here were formulated in such a way that mention is always made of the specific equation to which they relate, or to the kind of detailed description of the model that is deemed necessary.

In the same spirit, the last but one director of the CPB, Henk Don, in an article written in 2006 with Johan Verbruggen, then head of the CPB's business-cycle research department, wrote:

> Many policy measures in the macroeconomic sphere can only be understood and discussed properly with the help of a model which sets out the key relationships between macroeconomic variables. Such a model is an important instrument in considering relevant relationships.
>
> (Don and Verbruggen, 2006)

Tinbergen's model formulated a closed structure of the Dutch economy, within which it was possible to ask specific questions. The model refused to admit questions that could not be answered on its own terms, and that, as Don and Verbruggen make clear, is a view still held by the CPB.

How a model fixes the parameters of relevant policy questions is expressed rather well in the following visualization of the structure of the model by J.J. Polak, a colleague of Tinbergen's at the Handelshogeschool in Rotterdam, today's Erasmus University, where Tinbergen was given a professorship (see Figure 4.12). It is clear that a fixed structure of economic variables reproduces itself through time, such that the values of the variables are determined by data that lie outside the economic system. Polak stressed that economic theory indicates how the arrows between the variables needed to be drawn. This meant that the model functioned as a neutral platform by expressing political issues in quantitative terms.

58 *Business-cycle research*

It structured the question and streamlined the answer, while at the same time making it impossible to pose questions about the structure itself. The structure of the economy obeyed the timeless laws of the pendulum, and specific quantitative estimates could be made based on the time series collected by the CBS and other institutions.

When the CPB was established after the Second World War, with Tinbergen as its first director and a model inherited from the pre-war years, it found itself with the task of umpire in the political debate. To this day, the CPB calculates what the effects would be of the implementation of election manifestos, across the spectrum from left to right, as well as regularly examining the economic consequences of cabinet plans. Joop den Uyl, leader of Partij van de Arbeid (the Dutch Labour Party) in the 1970s, regularly talked about the narrow margins of democracy. It is no exaggeration to say that in the Netherlands, these narrow margins are still to this day determined to an important extent by the framework of the CPB models.

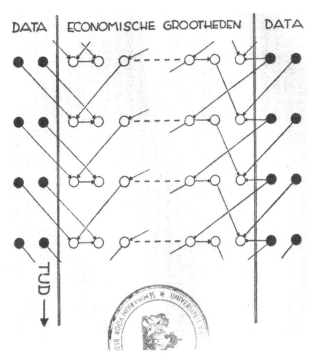

Figure 4.12 J.J. Polak's representation of the relationship between the structure of a model and its data.

Source: Adrienne van den Bogaard (1998). *Configuring the Economy: The Emergence of a Modelling Practice in the Netherlands, 1920–1955.* (Amsterdam: Thela Thesis), p. 126.

4.7 In conclusion

In this chapter, we have discussed the transition from a statistical to a model-based representation of an economy by looking at the business-cycle research carried out at the CBS between the wars by De Bosch Kemper on the one hand and Tinbergen on the other. Like Jevons in Chapter 3, Tinbergen was a hybrid expert; his personal enthusiasms and his training enabled him to cross the boundaries between disciplines.

Such a crossing of boundaries was characteristic of the new econometrics. In the first issue of *Econometrica*, the discipline was described as a combination of economic theory, mathematics and statistics. Coming from the natural sciences, Tinbergen combined those three elements in an entirely novel way, in a model, a mathematical representation of an economy that, at least such was the intention, contained a dynamic mechanism with which economic events could be explained and to a certain extent predicted. But the model could also be used for the purpose of simulation, to calculate the consequences of policy scenarios.

This brings us to a second reason why Tinbergen can be described as a hybrid expert. Although he became a professor at Erasmus University at a young age, he did not win his academic spurs at the university (or the Handelshogeschool as it then was) but at civil-service institutes such as the CBS and the CPB, whose purpose it was to support policymakers. For the principled, socialist and pacifist Tinbergen, economic science clearly served a social purpose. The position he had in mind for the science of economics was that of umpire and mediator. It could not determine the aims of policy, but it could show what the consequences of intended policies would be and indicate the means that would best serve to achieve a stated goal.

The economist thereby took up a position outside the glare of political controversy. Whereas political economists such as Mill and Cairnes had deliberately sought out public debate and saw their work as a political intervention, with Tinbergen's conception of the relationship between economics and politics, the economist withdrew into the wings. Mill's distinction between the science and the art of political economy prepared the way for this change, but it took a different method of working, that of an engineer or instrument-maker rather than a detective and playwright, to separate economics from politics. Jevons had already expressed a preference for the term 'economics' over 'political economy'. With Tinbergen's work, the separation became a reality.

References

Tinbergen, Jan (1926). Letter to Wibaut of 14 September 1926. *Wibaut Archives*. (IISG, Amsterdam) [original in Dutch].
Don, F. J. H. and J. P. Verbruggen (2006). Models and Methods for Economic Policy: 60 Years of Evolution at CPB. *Statistica Neerlandica*, 60(2), pp. 145–70, 146.

5 John Maynard Keynes and Jan Tinbergen
The dramatist and the model-builder

5.1 Introduction

In Chapter 4, we looked at the rise of a new way of carrying out economic research, the building of empirical economic models. To illustrate it, we examined the business-cycle research of the CBS and the work of Jan Tinbergen. We saw a shift from inductive statistics to an approach in which a model is constructed based on conjecture about the underlying mechanism that produces a picture matching the shape of the statistical data collected by the CBS. In his model, Tinbergen tried to give shape to this mechanism with the help of insights provided by economic theory. He then produced the equations in his model based on statistical data. This use of statistical data might be seen as a test for the model, but soon one of the most important matters under debate was precisely the degree to which Tinbergen's econometric evaluations could be regarded as a test. In this chapter, we shall look at John Maynard Keynes' criticism of Tinbergen and see that Keynes regarded even Tinbergen's way of working as inductive statistics. We will examine Keynes' criticism against the background of his own approach to economics, which was still indebted to that of his nineteenth-century British predecessors. In Chapter 6, we shall take a more detailed look at Milton Friedman's ideas about the testing of econometric models, and about the testing of economic theory more generally.

The vocabulary of evaluation and testing that was introduced with the rise of econometrics changed the playing field and the character of the discipline of economics forever. In earlier chapters, we saw that there had already been moves to identify statistical data as empirical evidence, as illustrated by the work of scientifically orientated economists such as William Stanley Jevons. From this point on, economic theory becomes mathematically formulated and needs to be evaluated and tested against statistical datasets. The playing field of economics as a discipline is described in mathematical, quantitative and statistical terms – anything that cannot be described in such terms no longer belongs in the science of economics.

This new vocabulary brought economics far closer to the types of problem that the philosophy of science highlighted in the twentieth century. The science of economics was no longer any different in character from other sciences. Like scientists in other disciplines, economists formulate hypotheses that then have to be tested. The problems faced by economists carrying out such tests have to do with the feasibility or otherwise of gaining sufficient control to ensure valid test conditions and are therefore no different in essence from those faced by meteorologists or astronomers who – such was the argument for many years – likewise do not have the option of carrying out controlled experiments. Yet neither the economist Lionel Robbins, whom we met in Chapter 3, nor John Maynard Keynes felt at ease with this conception of economics as a discipline. Hence, the importance of looking at the turf wars between Keynes and Tinbergen, which were firmly settled in favour of the model-builders.

5.2 John Maynard Keynes and the *mis-en-scène* of economics

John Maynard Keynes (1883–1946) was unquestionably one of the most influential economists of the twentieth century. He made his greatest impact, both politically and in the economic realm, through his membership of the British governmental delegation that took part in negotiations over German reparations after the First World War and through his important role as negotiator on the reorganization of the international monetary system as defined by the Bretton Woods agreements of 1944. Those negotiations damaged his already precarious health to such an extent that he died two years later. The Bretton Woods arrangements involved a system of fixed exchange rates with the dollar as the key currency, and the setting up of two international monetary institutions, the International Monetary Fund (IMF) and the World Bank. The IMF was intended to fulfil the same function for the international financial world as the national central banks for each country, namely that of lender of last resort, while the World Bank, with the end of the Second World War in sight, concentrated on efforts to rebuild, especially in what later became known as the Third World. These agreements and the accompanying institutions took the place of the 'gold standard', which was seen by many, including Keynes, as partly responsible for the deep economic depression of the 1930s.

The Bretton Woods System survived until 1971, when American president Richard Nixon, faced with rampant inflation caused by the Vietnam War (and speculation against the dollar), felt forced to end the link with gold that had helped to secure the dollar's position as the world's reserve currency and an international means of payment. It is important to note that Keynes had always been an opponent of the dollar's use as a global currency, and during the conference he unsuccessfully advocated the creation of a supranational currency (the 'Bancor') that would be governed by the IMF – a proposal heard again since the current financial crisis began in 2008. With victory in sight and with Britain virtually bankrupt, his proposal was impossible for the American delegation to contemplate; it was clear that after the war, the United States would want to use its military and political dominance to increase its economic power.

It was only after the Second World War that Keynes' name became synonymous with the IS–LM model and with a particular type of economics: demand-regulating, counter-cyclical budgetary policy. Much maligned in the last two decades of the twentieth century, this policy has returned to favour in some circles of late. The IS–LM model, which describes the interplay between money and production in an economy, is still the model used to familiarize first-year economics students all over the world with macro-economics. It shows how decisions about investment and savings (IS) act upon each other, while every economy also features a coordination of decision-making in the financial sphere between liquidity preference (or the demand for money) and the money supply (L and M respectively). Outcomes in these different markets need not necessarily coincide, and the decision made will certainly not automatically lead to a market outcome that produces full employment. It is astonishing to hear from one of the main critics of this model within the discipline of economics – Nobel Prize winner Robert Lucas – that the IS–LM model was used in education until the 1970s even in the Chicago of free-market economist Milton Friedman, Keynes' polar opposite (of whom more in Chapter 6).

It is because of the IS–LM model that Keynes is regarded as the man who first introduced macro-economic modelling into economics – a conclusion quite without justification. If that honour should go to anyone, then it is the Oxford economist Sir John Hicks who, with a few simple equations, tried to provide insight into the ideas contained in Keynes' magnum opus, *The General Theory of Employment, Interest and Money* (1936), which to many were unfathomable. After a deafening silence lasting six months, Keynes wrote to Hicks that he had 'next to nothing to say by way of criticism' of his model.

It would be quite wrong to conclude from this that Keynes was a builder of mathematical models himself. On the contrary, his long delay in responding to Hicks' model demonstrates his relative indifference to them. As we shall see, Keynes reacted with downright hostility to the mathematical–statistical approach of Jan Tinbergen. In a famous critique of Tinbergen's econometric study of business cycles, published by the League of Nations in 1939, Keynes ends by asking with a sigh whether at this particular juncture (1939) the League of Nations has nothing more pressing to think about.

Keynes distanced himself not only from economists such as Tinbergen but from his predecessors as well, labelling them classical – in other words: dated. But despite his scepticism regarding the likes of David Ricardo and John Stuart Mill, he remained strongly attached to the discursive approach that we examined in Chapter 3 in the context of John Elliot Cairnes' notes for Mill and his book *The Slave Power*. British political economists worked their notes up into coherent arguments. They recalled economic events in order to understand economics, editing their observations to create credible plots featuring credible economic actors.

5.3 Keynes' *The Economic Consequences of the Peace*

In what his biographer Lord Skidelsky calls perhaps his best book, *The Economic Consequences of the Peace* (1919), Keynes took precisely this approach. He had grown up in the sheltered environment of Victorian Cambridge. His father, John Neville Keynes, held a senior administrative post at the university, wrote a book about the method of political economy that remains very readable to this day but was too shy to set about building a proper academic career for himself. While a student, John Maynard Keynes was selected to join the Apostles. This secret student society and the circle of artists known as the Bloomsbury group were to be his two most important moral and aesthetic poles of reference for the rest of his life but Keynes' compass pointed in another direction too, as we shall see, not towards pure art but towards the world of politics and policy.

Both the Apostles and the Bloomsbury set strove for a form of moral perfection that expressed itself in a completely internalized aesthetic ideal. Their stress on the arts left no room for any political or public engagement. Before the First World War, Keynes had been interested mainly in Mediaeval Latin poetry, ethics and the mathematics of probability, and his circle of friends therefore regarded it as a form of betrayal when, just before war broke out, Keynes decided to leave his manuscript on probability to lie fallow for the time being and take a job at the Treasury. The devastating impact of the Great War on the British intellectual and governmental elite changed Keynes from a gifted boy who excelled in the entirely self-absorbed worlds of Eton and Cambridge into the most important and influential political economist of the first half of the twentieth century, certainly in Britain and perhaps in the world.

As a member of the British governmental delegation, Keynes took part in negotiations that culminated in the Treaty of Versailles of 1919 and saw from close proximity how talks about reparations were – as he saw it – pushing Central Europe towards the abyss. The outcome would be disastrous. In *The Economic Consequences of the Peace*, he expressed damning criticism of the Versailles Treaty and the reparations imposed on Germany, a country that had not even been truly defeated. In his angry and bestselling attack on the blind ambition of Allied politicians, Keynes placed his hopes in the science of economics, only to realize over time that the economic theory passed down to him did not provide the arguments that could give a more practical dimension to his outburst of rage.

Skidelsky's brief but apt description of the book enables us to understand how the mind of an economist such as Keynes actually worked. The Keynes who wrote it was not the economist he was to become; he was still the bright young thing who won all the prizes at Eton and Cambridge, always outstripped his opponents in arguments (and revelled in doing so) and felt morally committed to an ethereal form of ethics and to the Bloomsbury group. But in his career he had already started out on a different path.

Keynes' book is not simply a settling of scores with his years at the Treasury during one of the most dramatic periods of modern history, it is also a piece of self-justification aimed at his friends. Virginia Woolf, who had no faith at all in Keynes' aesthetic judgement, praised *Economic Consequences* highly. It has an engaging plot and, perhaps because it is not art but politics, Virginia Woolf felt able to approve.

The book begins in the nightmarish, theatrical setting of negotiations at Versailles, an experience Keynes said turned him from an Englishman into a European, and moves on through a description of the 'utopian' nineteenth century, in which the Malthusian population question seemed solved and prosperity for all within reach, to a detailed description and analysis of the Versailles talks, with the disastrous treaty as their final outcome. The text and the notes are saturated with statistical facts, sometimes serving to support Keynes' verdict on the pre-war socio-economic situation, sometimes designed to give a reasonable estimate of, for example, the sum that Germany, in the most favourable circumstances imaginable, would be able to repay.

Central to the book is a brilliant description of the personalities of the most important negotiators, George Clemenceau, Woodrow Wilson and – in passages much bowdlerized on the advice of his parents, since Keynes would have to continue his career in Britain – Lloyd George, the British prime minister. Keynes shows how the chemistry of these characters and a succession of events during the negotiations eventually led to the outcome he deplored. Keynes' message was clear. The negotiators had sacrificed long-term economic stability and peace in Europe, and therefore the world, for short-term political advantage. Clemenceau comes out of it all looking the most consistent of the three, since he alone kept a clear aim in view at all times, if the wrong one: to keep Germany in the stone age for at least a generation. Keynes believed this policy was unacceptable not only on economic grounds but more generally. It is worth quoting Keynes' words in full (1920, p. 225):

> I cannot leave this subject as though its just treatment wholly depended either on our own pledges or on economic facts. The policy of reducing Germany to servitude for a generation, of degrading the lives of millions of human beings, and of depriving a whole nation of happiness should be abhorrent and detestable – abhorrent and detestable,

even if it did not sow the decay of the whole civilized life of Europe. Some preach it in the name of Justice. In the great events of man's history, in the unwinding of the complex fates of nations Justice is not so simple. And if it were, nations are not authorized, by religion or natural morals, to visit on the children of their enemies, the misdoings of parents or of rulers.

In a famous letter to his friend and fellow economist Roy Harrod, Keynes wrote that as far as he was concerned, economics was a 'moral science'. It was inseparably bound up with an analysis of the motives of actors in the economy and could not escape the need to consider the complexity of economic life as a whole and the connections between economics, law and morality: 'I mentioned before that [economics] deals with introspection and with values. I might have added that it deals with motives, expectations, psychological uncertainties. One has to be constantly on guard against treating the material as constant and homogenous.'

Cairnes' book about the American slave economy had been in part a political and moral appeal to the British public and to British politicians, and Keynes' book on the consequences of Versailles had a similar purpose. Economics was political economy; it was not just an analysis of the economic state of affairs but an intervention in the social and political domain. This makes Keynes' concept of the practice of economics significantly different from that of Lionel Robbins, who had defended the separation of positive and normative economics, and from that of John Stuart Mill, who in his own way wanted to make a distinction between political economy as positive and as normative science. Keynes' ideas were imbued with historical detail. There was no clear dividing line between his work as a theoretician and his interventions in public affairs. Keynes' economics was political economy in the full sense of the term.

5.4 *The General Theory of Employment, Interest, and Money*

The Economic Consequences of the Peace lacked the economic theory that for a man like Cairnes in 1862 was self-evidently to hand in the work of David Ricardo and John Stuart Mill. The theory Keynes needed did not exist at that moment, in fact it would not arrive until the publication of his own *The General Theory of Employment, Interest, and Money* in 1936. There is a clear parallel between the two books. *Economic Consequences* tells a concrete story, involving real characters, while *General Theory* tells an abstract story about the workings of a market economy, involving abstract characters. Clemenceau, Wilson and Lloyd George were replaced by the stock-market speculator, the entrepreneur and the worker–consumer. Their 'animal spirits' took the place of the dubious motives of the politicians. Indeed, *General Theory* has a plot as well, made up of a succession of events that cause a market economy to reach a stable condition of mass unemployment without any automatic forces or mechanisms coming into play to lift it out of that condition.

Even though the characters in *General Theory* are abstract categories, Keynes' intentions in writing it were not purely theoretical. Like his book about the consequences of the Treaty of Versailles, it was written in response to his reflections on historical material that, in Keynes' own words, was not 'constant and homogeneous'. He merges the motives of actors and the changing institutional, economic and political circumstances into a coherent picture. Keynes saw his *General Theory* not as an abstract theory that would need to be tested against homogeneous statistical data but as bringing to the fore the underlying principles that he had extracted from his own rich experience and from the vastly diverse sources he consulted. Nor was it a neutral book. As with his work on the Treaty of Versailles, he wanted to influence the economic policy of the United Kingdom. 'Can Lloyd George do it?' was the question Keynes posed in 1929 in *The Nation and Athenaeum*: would the prime minister, by means of monetary inflationary policy, be able to finance Great Britain out of the crisis? Answering that question, like the writing of *General Theory*, was not simply a theoretical exercise. It was a contribution to the public debate.

Keynes elaborated on his economic 'cast of characters' in the famous Chapter 12 of *General Theory* and then, in the next chapter, outlined how, because of an incongruence of expectations, those characters found themselves at such an impasse that the market was on a dead-end track. Keynes formulated an alternative to the prevailing belief in the perfect operation of markets. As a result, he, a liberal, broke with the cherished doctrines of free trade and balanced budgets that had prevailed in Britain since Gladstone (and not only there, and not only then). After Keynes, political intervention in a market economy was no longer taboo; counter-cyclical budgetary policy became at the very least a meaningful idea, unemployment a failure of the market. Today's economists may think very differently about these matters, but that does not alter the fact that Keynes' *General Theory* set the agenda for economists and politicians for decades.

It is hard to imagine a more radical break with Victorian faith in the free market, balanced budgets and Gladstone. *General Theory* created almost tangible confusion among Keynes' contemporaries and fellow economists (indeed even among his closest students and colleagues, to whom he entrusted his manuscript). But despite the explosive content, his way of working was still that of the Victorian economist. He went in search of the right characters and the right plot, and in doing so he used the same diverse sources as his nineteenth-century predecessors, just as he had done in earlier work. The Victorian faith in the psychological plot was retained at an abstract level in Keynes' approach to economic theory.

On the very day *General Theory* was published, the Cambridge Art Theatre, which had been built largely with money donated by Keynes, opened for its first performance. That same month, Keynes' Russian-born wife, Lydia Lopokova, played Norma in Ibsen's *A Doll's House*. Two years earlier, Virginia Woolf had been enthralled by Lydia's performance as Nora in that same Ibsen play. With the best of intentions, without a hint of irony, she described Keynes being moved to tears by Lydia's acting:

> Dear Old Maynard was – and this is true – streaming tears ... I kissed him in the stalls between the acts; really she was a marvel, not only a light leaf in the wind, but edged, profound, and her English ... gave the right aroma.
>
> (Skidelsky, 2003, p. 513)

For Keynes, economic theory was the right story with the right characters, although not necessarily with a happy ending. Keynes' biographer Skidelsky clearly sees the theory as a narrative of sorts:

> The *General Theory* may be interpreted as a moral drama, but it is not a historical or a sociological one. The social landscape visible from Keynes' earlier writings has vanished: The Puritan rentier class, the business class, the workers, are represented only [as] disembodied 'propensities'.
>
> (Skidelsky, 2003, p. 513)

It is of course ironic that Keynes' name was associated for so long with one of the most influential economic models to be developed in the twentieth century, the IS–LM model. This might be interpreted as the unintended consequence of Keynes' own enthusiasm for those so-called business-cycle barometers that were developed in the early decades of the twentieth century to provide a picture of business cycles and, if possible, to predict them. In section 5.5, we shall see that Keynes' enthusiasm for such barometers did not translate into enthusiasm for the use of the models that were developed as a result of Jan Tinbergen's work. In fact, Keynes' fierce criticism of Tinbergen can be regarded as the swansong of a way of doing economics that was driven out of university economics departments after the Second World War. In this sense, the contrast between Keynes and Tinbergen is an example of the split between 'humanities' and 'sciences' that, in the 1950s, came to be seen as a clash of cultures. Keynes' approach to economics was replaced by the model-based approach of Hicks and Tinbergen.

5.5 Keynes' criticism of Tinbergen and econometrics

In 1936, Tinbergen left the CBS, having been commissioned by the League of Nations to research business cycles (see Figure 5.1). The results appeared in 1939 in two fat volumes that contained a similar model for the United States as that which Tinbergen had already developed for the Netherlands. The most notable (in Keynes' case ferocious) reactions came from two economists,

Keynes and Friedman. In 1939, Keynes was perhaps the most important living economist, while Friedman, who was carrying out statistical–empirical work for the American National Bureau for Economic Research (NBER) at the time, had yet to grow into one of the most prominent and influential economists of the post-war period. Friedman's criticism, which we shall examine more closely in Chapter 6, was mainly focused on the issue of whether the statistical methods of testing used by Tinbergen were adequate (he claimed they were not). For his part, Keynes mounted a frontal attack on Tinbergen's mathematical–statistical approach to economics.

Keynes expressed his criticism of the first part of Tinbergen's work for the League of Nations in a review published in *The Economic Journal*, of which Keynes was editor-in-chief. It is telling that Keynes took it upon himself to discuss Tinbergen's work in the house journal of the Royal Economic Society. Clearly he did not want to leave that particular task of demolition to anyone else. The entire tone of his review is condescending. Tinbergen, Keynes writes, sets to work so cautiously that ultimately he appears willing to concede 'that the results probably have no value'. It seems to Keynes that Tinbergen is concerned above all to get the job done, irrespective of the value of that job.

Figure 5.1 Professor Jan Tinbergen and the staff of the department of business-cycle research and mathematical statistics. The photograph was taken shortly before Tinbergen's departure in 1936 for Geneva, where he was to carry out business-cycle research for the League of Nations. Back row from left to right: A.L.G.H. Rombouts, P. de Wolff, J.B.D. Derksen, B. van de Meer, M. Eisma and K. Struik. Front row from left to right: M.J. de Bosch Kemper, J. Tinbergen and J.C. Witteveen.

Source: Courtesy of Beeldbank CBS.

Girded with rhetoric of this kind, Keynes arrived at his more substantial objections to Tinbergen's mathematical–statistical approach. He believed Tinbergen preferred statistics and arithmetic to logic, whereas he himself would have made precisely the opposite choice. An economist needed to have a complete picture of the economic situation and course of events before he could start building a model – that is, if he wanted his model to be of any use in making a quantitative estimate of the influence of the different variables. Still, the fact of the matter was that such an exercise was superfluous.

How independent and constant, after all, were the connections between economic variables through time? And did the stress on measurable variables, as produced by statistics, suggest that qualitative, 'non-statistical factors' were less relevant? Or even not relevant at all? What was the functional form of the links between variables? Were they all, as in Tinbergen's model, linear? Was this not a manifest and misleading simplification? According to Keynes, Tinbergen had simply dismissed all kinds of relevant matters. He left out of account all political, social and psychological factors, including governmental intervention, technological progress and expectations of the future, any of which might turn out to be of great importance to a relevant economic analysis. It was precisely the interplay between all these diverse factors that was so important in Keynes' own understanding of economics.

Keynes formulated several more technical objections to Tinbergen's business-cycles model as well. Tinbergen seemed to have introduced time lags into his variables by a process of trial and error, time lags that seemed to serve mainly as a way of making the equations in his model fit as closely as possible with the statistical time series he was using. The opportunism and the 'devastating inconsistencies' of most of these statistical series caused Keynes to fear the worst for the final result.

Keynes recognized that Tinbergen's book was a pioneering work, 'full of intelligence, ingenuity and candour', but the result, he claimed, was 'a nightmare to live with'. He feared that if Tinbergen were to agree with his objections, then he would simply hire yet more arithmeticians ('computers') and drown his sorrows in arithmetic. It was a mystery to Keynes why in 1939, on the eve of a new catastrophe, the League of Nations was devoting its time and money to an undertaking of this kind.

The tone and nature of Keynes' criticism was such that fellow econometricians rushed to Tinbergen's side. Jacob Marschak and Oskar Lange, director and fellow of the Cowles Commission respectively, went through what Keynes had said point by point to make clear how little Keynes had understood about Tinbergen's new way of doing economics. The Cowles Commission was the leading econometrics institute in the United States, and in those days it was attached to the University of Chicago. After the Second World War, it moved to Yale University and adopted a new name, the Cowles Foundation.

Marschak and Lange believed that the discipline of economics had no choice but to follow and improve the route laid out by Tinbergen 'if it want[ed] to be taken seriously'. Anyone placing their faith in Keynes' way of working would only be adding to the 'non-ending list of plausible essays'. Like Tinbergen, Marschak and Lange concentrated on improving and further systematizing the collection of statistical time series, which meant that as the years went by they were increasingly able to overcome Keynes' justified objections about the poor quality of the statistics used. But they resisted the emphasis Keynes placed on the combination of diverse kinds of observations. It was best, they believed, to make measurable and comparable as many as possible of the factors that Keynes claimed fell outside the domain of statistics, so that their influence could be evaluated.

As editor-in-chief of *The Economic Journal*, Keynes refused to publish this response by Marschak and Lange. It was a remarkable sign of weakness. To Keynes, the task of the economist was to weigh up different sources of information, using his judgement to sift the important connections out of all the statistical, literary and other sources he drew upon, discarding everything that was not of immediate significance. The economist's personal judgement was inextricably bound up with his analysis of the economy.

There was a crucial distinction here: between risk and uncertainty. Keynes saw risk as something that could be quantified, since it was subject to the laws of probability, but uncertainty could not be quantified in the same unambiguous way. It was down to the economist to weigh up information from various sources; there was no homogenous space in which the distribution of probability could be straightforwardly attributed to these diverse sources of information.

The new econometricians, by contrast, subscribed to the creed of one of the founders of modern statistics, Karl Pearson, who said that any serious academic must above all attempt to be objective in his research. The quality of an analysis depended on the mathematical rigour of the economic model and the quality of the statistical data series. Regression and correlation coefficients and not the virtuosity of the individual economist would henceforth determine the value of theoretical insights. With the rise of subjective probability theory, the distinction between risk and uncertainty that Keynes saw as fundamental was about to lose its significance. In modern probability theory (referred to as Bayesian), everyone is expected to attribute a subjective but unambiguous probability to equally unambiguous events.

5.6 A tale of two cultures

In 1959, C.P. Snow, author, senior civil servant and important British public intellectual, delivered his famous Rede Lecture in Cambridge about a fundamental divide between the humanities and the sciences that had arisen in the twentieth century. People in arts faculties were no longer able to understand science adequately, while scientists lacked the broad historical and social view of those engaged in the humanities. This had damaging consequences for society as a whole. Snow's much discussed distinction between arts and sciences seems directly relevant to the disagreement between Keynes and Tinbergen.

Keynes may have been a statesman and a socio-economic reformer but to the post-war way of thinking, he was no longer an economist. His working method was diametrically opposed to that of the scientist, who derived information from statistical data using the toolkit of the modern econometrician. Marschak and Lange's reproach was that with his long-winded style and complex references, Keynes produced only plausible stories, rather than testable hypotheses. Keynes' work might still be of value, but only as a theoretical or literary exercise. Mathematical theorizing and statistical testing promised more clarity and precision than Keynes' essay writing. Keynes' approach no longer fulfilled the conditions of empirical science.

In his biography, Robert Skidelsky draws a direct parallel between the conception of art held by the Bloomsbury set and the rise of mathematical economic models:

> [They] located beauty not in the subject matter or 'narrative' of a work of art, but in its formal structure, intuitively apprehended; the shift from flow of narrative to flow of thought is a distinguishing characteristic of Virginia Woolf's novels. A parallel shift towards formalisation, or model-building, was taking place in economics.
> (Skidelsky, 2003, p. 456)

Yet it was precisely this parallel tendency towards the mathematization of economics that Keynes rejected. Based on the work of Jevons, Edgeworth and Marshall, he had observed even in his student days that the prospects for a mathematical and statistical approach to economic issues were fading fast, since such an approach lacked the potential to bring into the open the complexity of economic processes and their intrinsic link to a specific time.

This was how he reacted to Hicks' mathematical model as well, with unease, because the complexity of the plot of his *General Theory* could not be found in it. As Keynes saw it, the model-based approach of Hicks and Tinbergen rendered up only poor economic analyses.

The prevailing view today is that although Milton Friedman had a negative attitude to Tinbergen's model-based and statistical exercise, he was constructive in his criticism, whereas Keynes' frontal attack on Tinbergen was primarily destructive. Keynes' criticism is therefore usually pushed aside as irrelevant, all the more so since Keynes hit out at Tinbergen even before he had read the second part of the study. But behind Keynes' criticism lay a substantially different conception of the character of economics as a science and the task of the economist that flowed from it. To him the proper practice of economics was to search for the right characters and the right plot. A good economist managed to distil this plot from the heterogeneous and complex multiplicity of empirical data and use the resulting analysis to intervene in the political debate.

After the Second World War, this kind of economics rapidly fell out of favour. Economists concentrated instead on refining mathematical theories and statistical techniques, which they then used to derive information from quantitative statistical data. Such efforts belonged to the realm of inductive statistics that Keynes regarded with such disdain. Keynes' refusal to publish the defence of Tinbergen by Marschak and Lange ultimately amounted to shooting himself in the foot. Keynes tried to ignore what Marschak and Lange clearly saw: real economists would have to devote themselves to developing economic models and to testing those models against statistics. In their view, Keynes was fighting a rearguard action; his disdain for their way of working was an expression of the literary virtuoso's misplaced sense of grandeur. A literary virtuoso could no longer be regarded as a scientist.

5.7 'Tinbergen's 2'

The difference between the two cultures identified by Snow is made wonderfully clear by the following anecdote. In interviews, Tinbergen never tired of telling the story of his only face-to-face meeting with Keynes:

> I met [Keynes] in person just once, shortly before his death, and I enjoyed that meeting greatly. His certainty regarding his vision of the issue of German reparations emerged in a humorous way, as follows. In that vision of the matter he assumed that a reduction in prices by an exporting country of 1% would result in an increase of 2% in the volume of its exports. No one had yet verified this. When I told him I had made an attempt and had indeed reached a figure of 2, I thought he would be pleased. His response, however, was merely: 'That's nice for you.'
>
> (Magnus and Morgan, 1987)

To this day, staff at the CPB will introduce this anecdote without any prompting, even though there is nothing funny about it since it was clearly a source of frustration to Tinbergen. In the various versions in which it is told, Tinbergen uses an almost identical form of words. At the CPB, the number 2 as a figure for export elasticity is therefore known as 'Tinbergen's 2' and Tinbergen never tired of stressing that it was actually Keynes' 2. Now bear in mind that in interviews, Tinbergen always took pride in saying that he never read books but instead had his wife read them to him. The relevant passage in Keynes' *The Economic Consequences of the Peace* goes as follows:

> Let us put our guess as high as we can without being foolish, and suppose that after a time Germany will be able, in spite of the reduction of her resources, her facilities, her markets, and her productive power, to increase her exports and diminish her imports so as to improve her trade balance altogether by $500,000,000 annually, measured in pre-war prices. This adjustment is first required to liquidate the adverse trade balance, which in the five years before the war averaged $370,000,000; but we will assume that allowing for this, she is left with a favorable trade balance of $250,000,000 a year. *Doubling this* to allow for the rise in pre-war prices, we have a figure of $500,000,000. Having regard to the political, social, and human factors, as well as to the purely economic, I doubt if Germany could be made to pay this sum annually over a period of 30 years; but it would not be foolish to assert or hope that she could.

In the entire book, this is the only place where Keynes writes of a doubling in connection with exports and prices, and in this passage Keynes is clearly referring to a correction of the value of the trade balance (exports minus imports) to take account of pre-war prices, rather than the response of exports or imports to a change in price of 1 per cent. That would at the very least explain the rather curt reaction by Keynes to the result calculated by Tinbergen.

It is more probable that Keynes had no idea what Tinbergen was talking about. Tinbergen in turn must have misunderstood what his wife read out to him but, because he was more interested in numbers than in books, his misconception went on to live a life of its own. Keynes regarded his numerical example as a 'reasoned guess', not a factual estimate. He was using it to construct an argument, not to test a theory against empirical, statistical data.

The staff at the CPB were therefore quite right to talk of 'Tinbergen's 2' rather than 'Keynes' 2'. In Tinbergen's memory, Keynes' complex argument could be reduced to that with which he felt most at home: a number. This also fitted perfectly with the number-crunching culture of the CPB. To Keynes, that culture was simply 'a nightmare to live with'. Victorian political economists, including Keynes, were involved in an inductive quest in which they constructed a coherent plot based on a multiplicity of diverse sources. From the econometric revolution onwards, induction was identified with the application, as efficiently as possible, of statistical techniques so as to extract the maximum information out of given statistical datasets. The theories that Keynes derived from his inductive quest were replaced by mathematically formulated hypotheses that had to be tested against statistics.

Figure 5.2 John Maynard Keynes in his study in Bloomsbury, a photograph published as part of a *Picture Post* item on his plans for financing the war against Germany, 16 March 1940.

Source: Courtesy of Getty Images.

The difference between these two cultures of economic practice, between Keynes and Tinbergen, emerges quite strikingly if we compare the no doubt carefully posed photograph of Keynes in his study in Bloomsbury (Figure 5.2) with the equally posed photograph of Tinbergen among the staff at the business-cycles research department of the CBS (see Figure 5.1). In contrast to Tinbergen, Keynes is portrayed as the sovereign individual, in splendid isolation on his chaise longue, making notes in peace and quiet. He is surrounded by papers and a large washing basket that serves as a waste-paper bin. The telephone is within reach, in case a senior government official should call for advice. He is holding a pen and notebook in which, by deploying his characteristic erudition and style, he transforms his vast range of personal experience as a negotiator into a coherent picture, a plan for the successful financing of the new war against the Germans. His judgement is central. Keynes the economist is at the same time a literary figure, a moral hero. The caption supports this impression, reminding readers that Keynes, years earlier, angrily walked away from the negotiating table and by doing so passed up the chance of a senior position, since he foresaw the disastrous consequences of the Treaty of Versailles.

Tinbergen, although (rather uncomfortably) at the centre of the picture, is surrounded by colleagues who contribute to his model as a team. The model of the Dutch economy was not Tinbergen's alone, it was the result of collaborative labour, just as Tinbergen could not himself collect all the necessary data that he needed to construct his model at the CBS. The new empirical science of economics was no longer the work of individuals but of entire institutions, no longer discursive but mathematical and statistical.

References

Magnus, Jan R. and Mary S. Morgan (1987). The ET Interview: Professor Jan Tinbergen. *Economic Theory*, 3, pp. 117–42, 130.

Skidelsky, Robert (2003). *John Maynard Keynes 1883–1946: Economist, Philosopher, Statesman.* (London: Macmillan).

6 Milton Friedman and the Cowles Commission for Econometric Research

Structural models and 'as if' methodology

6.1 Introduction

The change to the playing field and character of the discipline of economics had one quite literal aspect to it. After the Second World War, the centre of gravity in economics moved to the United States, where it has remained ever since. This shift had begun in the 1930s with the emigration of academics and intellectuals, a large proportion of them Jewish, especially from Central Europe. Many of these exiles contributed to the rise of mathematical economics and econometrics in the United States. In the American context, the debate about the merits of econometrics took a different turn from that seen in Europe. The argument was specifically about the merits of empirical statistical research as against the theory-based modelling that came to fruition with the work of the Cowles Commission for Econometric Research. It was not about the merits of empirical statistical research as such.

Even before the Second World War, one major strand of research in the United States was focused on statistics, an example being the Economic Service at Harvard University, which, as we have seen, constructed business-cycle barometers. The years between the wars saw the birth of a further two statistically oriented research institutes: the NBER allied to Columbia University in New York and the Cowles Commission for Econometric Research. The Cowles Commission was to grow into one of the leading institutes for econometric research in the United States. Founded in 1932 by businessman and economist Alfred Cowles, it set up shop in Chicago in the university's department of economic and other social sciences, before moving to Yale University in New Haven in 1955. The Cowles Commission had a tense relationship with both the NBER and the economics faculty at Chicago. It is time to take a close look at the famous 'measurement without theory' controversy, which erupted shortly after the Second World War. Tjalling Koopmans, an economist and physicist of Dutch origin, and successor to Jacob Marschak as director of the Cowles Commission, expressed blunt criticism of an important study of business cycles by the director of the NBER, Wesley Clair

Mitchell, and his colleague Arthur Burns. Shortly after that, the econometric models developed by Lawrence Klein at the Cowles Commission came under intense scrutiny.

The source of this latter expostulation was Milton Friedman, who, as a former student of Burns and a member of staff at the economics faculty in Chicago, had close links with the NBER. Even before the war, Friedman had expressed restrained criticism of the study into business cycles carried out by Tinbergen for the League of Nations, demonstrating his profound interest in empirical statistical research. In essence, Friedman raised the question of whether Tinbergen's econometric model had actually been tested against statistical data, or whether it had, in fact, merely been adjusted as far as possible to fit the statistical data available. Unlike Keynes, Friedman worded his criticism in such a way that it suited the new playing field of the science of economics, where the boundaries are determined by mathematics and statistics. His assessment also featured a specific idea about the relationship between economic science and economic politics, a notion he expressed explicitly in his famous essay 'The Methodology of Positive Economics' of 1953. A discussion of Friedman's criticism of Tinbergen and the Cowles Commission therefore serves as a prelude to a review of that essay which, along with John Stuart Mill's essay of 1836, is among the most discussed, quoted and criticized articles on the methodology of economics.

At first sight, Friedman seems to defend the view that the structure of economics – which renders up the test criteria – is no different from that of other sciences. But his defence rests on an important assumption about the underlying image of science. According to modern philosophy of science, experimental science epitomizes the image of the sciences in general, but Friedman argued that economics was not compatible with that image. He also claimed that the use of 'the experimental method' as a reference point in identifying good science actually created a false impression of the character of the natural sciences. This gave rise to Friedman's famous (or notorious) 'as if' methodology: economic models do not represent reality, rather they are useful instruments based on largely unrealistic assumptions. This sixteen-word summary of Friedman's position relies upon notions that have been furiously contested and subjected to minute philosophical scrutiny. What is a model? What is an economic instrument? What does it mean to say that an assumption is 'unrealistic'? We shall see that Friedman's methodology contains a strongly ideological message about the relationship between economic theory, empiricism and economic policy.

6.2 'Measurement without theory'

Burns and Mitchell's *Measuring Business Cycles* (1946) was the product of many years of collaboration during which the NBER had meticulously gathered statistical data on economic trends. Based on their extensive datasets, they attempted to draw general conclusions about business cycles, and in that sense their book fitted within the inductive tradition of the NBER. Mitchell was the first director of the NBER, Burns his successor. Arthur Burns had studied under Mitchell at Columbia University. In the 1970s, during the presidency of Richard Nixon, as a result of his directorship of the NBER, he was appointed chair of the FED, the American system of central banks. He consequently became one of those blamed for the collapse of the Bretton Woods system in 1971. As for Milton Friedman, he was a student of Burns at Rutgers University in New Jersey. Later, at the University of Chicago, Friedman developed into one of the most important representatives of what came to be known as the Chicago School, the tightly knit economics faculty famous above all for its outspoken ideas about the beneficial workings of the market.

In Chapter 5, we saw that Keynes reproached Tinbergen for indulging in 'inductive statistics', meaning that Tinbergen had arrived at his econometric models based on statistical time series. That was true only in part. As became clear in Chapter 4, Tinbergen was looking for a theoretically informed mechanism, which he then compared with the available empirical time series. It was not a matter of one-way traffic from statistical data to mathematical model, since there was also movement in the opposite direction, from theory to data. Tinbergen's models functioned as a kind of lens, making it possible to connect theoretical notions with empirical facts.

The reproach of inductive statistics would have been more appropriate in the case of the work of Burns and Mitchell. Right from his earliest studies into business fluctuations, Mitchell proposed a programme in which he would ascertain the characteristics of business cycles by looking at empirical statistical time series. He believed that theories must be used opportunistically to explain the characteristics that emerged. The measurement of business cycles was his main objective, not the explanation of them. This was the programme that Burns and Mitchell followed in their joint study of 1946.

The outcome they were hoping for was a reference cycle, which would serve as a basis for comparison with other cycles. Differences between the reference cycle and a specific cycle could then be sought in specific explanatory circumstances. So the approach of Burns and Mitchell seems in keeping with the distinction made by John Stuart Mill between tendencies and 'disturbing causes'. But Burns and Mitchell saw their reference cycle not as a tendency in the sense of a general pattern but as an actual reference point, a basis for comparison, that would help them in seeking ways to explain concrete, specific business cycles. To create their reference cycle, they drew upon all kinds of statistical methods: moving averages were calculated, linear regressions were carried out and so on, all with the aim of arriving at a generic profile of a business cycle against which specific business cycles could be compared.

The result of their efforts can be seen in Figure 6.1. The difficulties in interpreting this and the many other figures in Burns and Mitchell's study were legion, not merely because of the unusual and novel character of their work but because of the *ad hoc* impression made by the methods they used and the *ad hoc* character of the explanations they offered. In earlier work, Mitchell had explicitly defended an opportunistic attitude to theory – take whatever you can use – and complained about a lack of interest among economists in making precise measurements of the phenomena they were studying. Now it was said of Burns and Mitchell that although they did try to measure, the status of their measurements was unclear and there was a lack of any theoretical foundation, which made those measurements difficult if not impossible to judge on their merits. The arbitrary nature of the methods and theories deployed meant that a reference cycle such as the one shown in Figure 6.1 was more a shot in the dark than a serviceable research tool.

This criticism was expressed most forcefully by the economist Tjalling Koopmans. Like Tinbergen, he had been trained in the natural sciences and was greatly interested in applying his talents as a scientist to the study of economics. In 1938, Koopmans left his native Holland for Geneva to become Tinbergen's collaborator in business-cycle research at the League of Nations. When the Second World War broke out, he moved to the United States. In July 1948, he took over from Jacob Marschak as director of the Cowles Commission. Under Koopman's leadership, the organization moved from Chicago to Yale in 1955, thereby bringing a definitive end to a relationship that was generally seen as troubled between the economics faculty at Chicago University and the Cowles Commission.

Written on the facade of the social sciences building in Chicago were the words (attributed to Lord Kelvin): 'If you cannot measure, your knowledge is meager and unsatisfactory', but this did not mean that all measurement was knowledge. Koopmans expressed his criticism of the inductive empirical statistical programme of Burns and Mitchell's *Measuring Business Cycles* (1946) in a famous review entitled 'Measurement without Theory', originally published in *The Review of Economic Statistics* in 1947. Koopmans set his sights high. He drew a comparison between what he called the 'Kepler' and the 'Newton' stages of science. Kepler, Koopmans argued, made use of very precise measurements by Danish astronomer Tycho Brahe to discover that the path of Mars around the sun is an ellipse rather than a (possibly rather egg-shaped) circle. But Kepler could only describe that path; he had no explanation for it. With his general law of gravity, Newton provided an explanation. Koopmans' argument was that whereas Kepler achieved merely a summary of the data, Newton offered insight into the underlying, explicatory mechanism. This is not the place to criticize Koopmans' assessment of Kepler's accomplishment, which does not withstand close scrutiny. We will simply take Koopmans' distinction between theory and data, and the pathways of inference between the two, as our point of departure.[1]

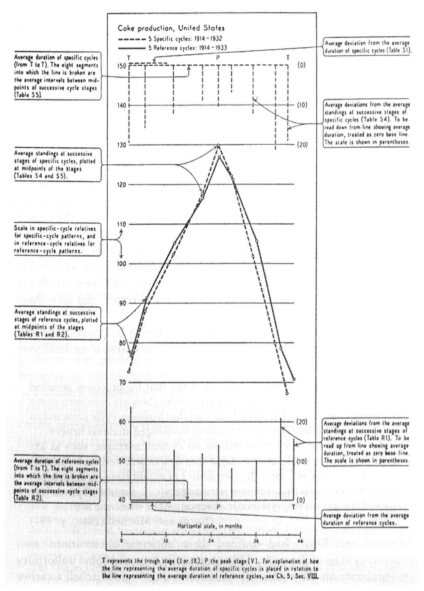

Figure 6.1 Mitchell's specific and reference cycle chart, in Burns and Mitchell, *Measuring Business Cycles* (1946).

Source: Chart 2, Sample Chart of Cyclical Patterns, in Arthur F. Burns and Wesley C. Mitchell (1946). *Measuring Business Cycles*. (National Bureau of Economic Research), p. 35.

In essence, we see here the same basic shift in ideas that Tinbergen had brought about in practice at the CBS. De Bosch Kemper's graphs merely summarized business cycles, whereas Tinbergen's model provided an explanation. This was an effective way of demolishing the work of Burns and Mitchell and, at the same time, putting the value to economics of his own Cowles Commission on a par with the significance of Newton for physics. More specifically, he was addressing the difference between a statistical description and a theoretical (causal) explanation of business cycles.

In Koopmans' view, the work of Burns and Mitchell was invalid on both fronts. He accused them of making use of a broad range of statistical techniques without asking themselves whether those techniques were compatible with each other or whether the statistical data to which the techniques were applied ought actually to be handled in such a way. They had made all kinds of implicit assumptions about underlying probability distributions but nowhere were those assumptions made explicit. Insofar as they provided 'explanations', it was unclear how their explanatory variables caused cyclical movements in the data. Their work was, in short, Koopmans said, a mishmash of descriptions, randomly chosen statistical techniques and uncorrelated explanations.

The econometric models developed by the Cowles Commission were a different matter altogether, Koopmans claimed. During the Second World War, partly in imitation of Tinbergen but also, importantly, building on the work of Norwegian econometricians Ragnar Frisch (under whom Koopmans had studied for several months) and Trygve Magnus Haavelmo, the Cowles Commission had formulated what were known as structural models: theoretically based models that described the causal structure of an economy. These models consist of behavioural and definitional equations with endogenous and exogenous variables. The idea was that the values of the exogenous variables were determined outside the model. To give an everyday example: the amount of sunlight may well influence the harvest but the reverse is not the case. Sunlight is therefore an exogenous variable, the harvest an endogenous variable. Endogenous variables are determined within the model. Parameters indicate which variables are dependent on each other and in what way (three-quarters of every additional euro of income is spent, for example).

Such models came to be known as structural models because the behavioural equations were deemed to describe the underlying, invariant, which is to say unchanging, causal connection between all relevant variables of an economy. A change to the interest rate (itself an endogenous variable) might, for example, have an effect on investment but also on savings. A change in savings, at a given level of income, affects consumption. Income then changes as a result and so on. The thinking behind such structural models is expressed in the system devised by Tinbergen's colleague Polak which we looked at earlier (see Figure 4.12).

The structural equations in the model were produced based on specific statistical time series. Building on the work of Trygve Magnus Haavelmo, the data in such time series were seen as the outcome of a natural experiment. 'Nature' drew from a large urn full of possible outcomes for (to take one example) the national income for a specific year. By regarding time-series data, such as the national income or national consumption, as random draws from an urn, it became possible to apply the complex instruments of probability theory to time series. As Mary Morgan has shown in her *History of Econometric Ideas* (1990) and elsewhere, the introduction of modern probability theory into econometric testing was revolutionary for the discipline of economics.

The importance of this step is indeed impossible to overstate. Before it, economists and statisticians had wrestled with the question of how probability theory, used so successfully in other fields of science, could be applied to the time series that were so important in the social sciences. Time series are unique, one-off series of data. Seeing them as pulled out of an urn (and therefore subject to a distribution of probability) made it possible to apply probability theory. Any differences between the estimates in the model and the statistical data were seen as 'noise' or 'errors'. Assuming that the data had a normal distribution, the errors or noise would average out to zero. (If that was not the case, the reason might be a hidden explanatory variable.)

In practice, it proved extremely difficult if not impossible to generate the structural equations of the model directly. The variables were after all interconnected. The question therefore was how you could be certain that you had indeed isolated the invariant structure of an economy. Look for example at the graph of simple supply and demand shown in Figure 6.2. The horizontal axis gives the amount, the vertical axis the price of a good (in this case maize). The only data we have that can be regarded as 'empirical' are statistical data about sales and prices. Those data are the points between the two axes. What a statistician (or economist) does not have are data about demand and supply separately. In principle, an infinite number of models can be constructed with the help of the given data. You know whether you have a supply function or a demand function only if one of the two is given a clear 'kick'. This is known as the identification problem and it is a fundamental difficulty for the builders of structural models in econometrics.

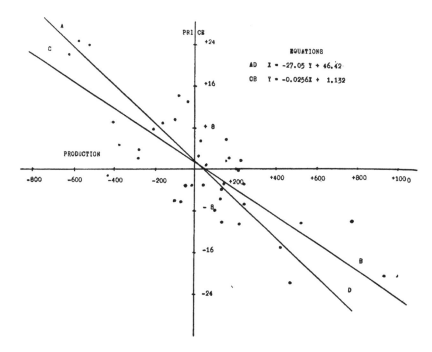

Figure 6.2 Scatter plot, representing maize production and price differences.

Source: W.M. Persons (1995). Correlation of Economic Statistics (1910), reproduced in David F. Hendry and Mary S. Morgan, *The Foundation of Econometric Analysis*. (Cambridge: Cambridge University Press), p. 138.

Think back to Stanley Jevons' graph of prices and quantities that we looked at in Chapter 3 (see Figure 3.2). We saw a number of points on his graph and because of the dotted line, it looked as if there was a demand function (if the price rises, demand declines). But every actual point on the graph is a result of supply and demand. Assuming that only the supply function has been given a 'kick' on a number of occasions, we could regard the dotted line as an estimate of a stable (invariant) demand function. This was the method used by American statistician and economist Henry L. Moore to estimate a demand function based on production and price statistics for maize in 1919 (see Figure 6.3); harvests might be good or bad, but the demand function would remain unchanged.

The idea of the Cowles Commission was that its econometric models reflected the underlying structure of an economy or, in other words, contained the economic mechanism that Tinbergen, too, was looking for in his model-based approach to business cycles. If we set aside the Cowles Commission's systematic use of probability theory, the quest for the underlying 'stable' explanatory mechanism of an economy linked the Commission's work to that of Tinbergen. A structural model was a representation of the causal structure of an economy.

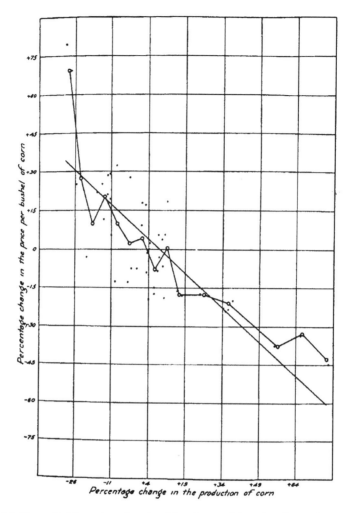

Figure 6.3 Henry L. Moore's estimate of the demand function for maize.

Source: Henry L. Moore (1914). *Economic Cycles: Their Laws and Causes*. (New York: Macmillan), p. 71.

Koopmans stressed that the econometric models produced by the Cowles Commission were theoretically motivated, unlike the statistical work of Burns and Mitchell. But in his original criticism of Burns and Mitchell, it remained unclear what theory the Cowles economists were using as the starting point for their work, and it was equally unclear on what grounds economists could have faith in the correctness of that theory. Koopmans pointed to the economic theory of the firm, which saw businesses as maximizing profit and consumers as maximizing utility. Between the lines, he suggested that the correctness of this theory could be shown by means of questionnaires and by introspection, a point to which we shall return later in this chapter. What are known as Walrasian equilibrium models, later to become of such importance to the work of the Cowles Commission, did not have any important part to play at this stage.

In Walrasian equilibrium models, an economy is conceived as a system of markets. A disturbance of the equilibrium in one market leads to a process of adjustment that influences other markets and will end only when a new equilibrium has been reached by all markets jointly. Such a model was first formulated by the nineteenth-century French economist and engineer Léon Walras in 1874.

These Walrasian models were to become increasingly important in the work of the Cowles Commission from the mid-1950s onwards, so important in fact that the staff at Cowles increasingly lost themselves in the mathematical problems that accompanied them, and their empirical econometric research gradually faded into the background. Mathematically, it turned out to be extraordinarily difficult to find a sound collection of conditions such that a process of adjustment leading to an overall balance could be observed.

At the point when Koopmans wrote his criticism, all this still lay in the future. His criticism was based on the firmly expressed conviction that, contrary to the views of Burns and Mitchell, theoretically informed but empirically oriented econometric research could achieve a systematic description and analysis of the structure of an economy.

In the early 1950s, the Cowles Commission finally nailed its colours to the mast by holding a conference dedicated to a 'test' of its showpiece model of the American economy, constructed by Lawrence Klein (who received the Nobel Prize for it in 1980). From various sides, the Cowles Commission was praised and criticized as a result. The most interesting of this praise and criticism came from Milton Friedman. Before we turn to Friedman, it is useful to take a quick look at Klein's model and the tests to which it was subjected.

6.3 A test of Klein's model

Lawrence Klein's model consisted of sixteen equations that described, for example, supply and demand in the case of investments, consumer goods and work. First, the model was constructed based on available time series for the pre-war period. Then the model was used to predict the values of endogenous variables for the years 1947 and 1948. The reliability of these predictions was assessed based on a number of different criteria.

There were several statistical criteria or tests that had to do with the consistency of the predictions and the question of whether the predicted values fell within certain predetermined margins. These statistical criteria are often used today as the only test of models. Think for example of indications of a model's estimated parameters, confidence intervals and so forth, elements that with today's statistical software can be generated quite simply. The key question is what a high statistical correlation between different variables (measured by one or other statistical criterion) tells us taken by itself. The Cowles Commission was fully aware of the importance of this question, which is why it compared the model predictions with the predictions of two so-called 'naive' models. The first naive model set the prediction for the coming year as equal to the actual value for the current year. The second naive model set the predicted value for the coming period as equal to the value for the current period, corrected to take account of a change in the current value in relation to the value in the previous period. As equations, these look as follows:

Naive model I:
$Y^*(t+1) = Y(t) + \varepsilon_1(t)$.

Naive model II:
$Y^*(t+1) = Y(t) + (Y(t) - Y(t-1)) + \varepsilon_2(t)$

$Y^*(t+1)$ stands for the predicted value of variable Y in period (year) t+1, while ε_1 and ε_2 are terms for random disturbance. We will now look in more detail at the importance of both naive models.

6.4 Was the test of Klein's model a success or a failure?

At the conference that the Cowles Commission devoted to the model, Carl Christ, who worked at Cowles, discussed the procedure that had been followed and the results of the test. The results varied. In the first statistical type of test, the model generally did well, with results that fell within the margin of error over the period for which the model had earlier been constructed, although this was not the case to the same degree for all the separate equations. The comparison with the naive models was far less encouraging. In more than half of the cases, both naive models predicted the endogenous variables better than the complex economic structural model did. (We should note that a comparison between the CPB model and naive models carried out in the 1990s likewise showed the naive models as producing better results.) The conclusion could only be that the Klein model was unable to withstand such a comparison. It is worth quoting the words used by Carl Christ in his conclusion:

> The econometric model used here has failed, at least in our sample consisting of the one year 1948, to be a better predicting device than the incomparably cheaper naive model, even though almost every structural equation performs as well, i.e., has just as small an error, in extrapolation to 1948 as it does in the sample period.
> (Christ, 1951, p. 80)

Nonetheless, Christ added that the econometric model was still to be preferred to the naive models, because unlike them the econometric model offered an insight into the causal structure of the American economy. The econometric model would therefore be able to indicate the consequences of various policy measures, or of changes in exogenous variables, whereas the naive models could say nothing at all about such things. The only conclusion to be drawn from naive models was 'that there will be no effect'. According to Christ, this counted against naive models particularly in circumstances of abrupt changes in an economy (for example, a sudden reversal in economic fortunes or a crisis resulting from the outbreak of war or from some other shock).

Christ was hereby suggesting that models should be used in the ways that Tinbergen had used his model in the interwar period, in his Labour Plan and in his policy memorandums as a civil servant. This is the way in which the CPB uses econometric models in the Netherlands to this day, to calculate what the consequences of policy measures will be. The practical consequences of its use are considerable; it is by relying on such model-based predictions that a government decides whether or not a significant restructuring of state finances is needed. Naive models, which merely say that today's weather will be like yesterday's weather or is following a specific trend, cannot be used in this way because they contain no explanatory variables. If the weather suddenly changes, a naive model will be more likely to produce the wrong results – is the thinking – than a model that demonstrates structural, causal coherence. Today, the CPB deploys precisely the same arguments as Carl Christ in defending its models.

So although Christ was certainly disappointed by the predictive performance of the econometric model in comparison to the naive models, he nevertheless expressed a preference for the econometric model because of the ways it could be used for purposes of policymaking.

6.5 Milton Friedman's criticism of the Cowles Commission

Christ's sketch and evaluation of the Klein model was the opening contribution at a congress entirely devoted to the Klein model. The first, extremely critical commentary on that model came from Milton Friedman.

Even before this, in a review of Tinbergen's work for the League of Nations, Friedman had argued that an econometric model should not just be constructed but actually tested. In that review, he had quoted with approval Wesley Clair Mitchell, who claimed that a 'competent statistician' with sufficient time and technical support would always be able to adjust a time series to a system of equations such that the correlation coefficient was above 0.9. High correlation coefficients therefore say nothing about the quality of a model, which can be read only from an independent test, which is to say a comparison with data other than those used for the estimates contained in the model. Friedman's criticism of Tinbergen was that he had failed to carry out such a test. Later in his review, he referred approvingly to the as yet unpublished work of Burns and Mitchell (which was panned by Koopmans), specifically regarding Tinbergen's use of annual figures when a far more refined analysis would have been possible if quarterly or monthly figures had been used. Friedman's review was friendly but firm in tone and it did not disguise his tendency to support his empirical teachers, Burns and Mitchell.

We see the same friendly firmness in his criticism of Carl Christ. First, he complimented the Cowles Commission on laying itself so open to criticism, but he then went on to make clear that he could follow little of the reasoning used by Christ to explain his preference for the Klein model as opposed to the naive models. At this point, Friedman was emphasizing not the importance of an independent test using a separate dataset but rather the comparison of the Klein model with the two naive models. A real test, Friedman argued, did not just assess the internal qualities of a model, it compared a model with a different, alternative model. That was precisely the function the naive models were intended to perform, and the conclusion could only be that the econometric model was not so good and was therefore unsuitable for other purposes as well.

Christ had argued that the complex structural model was suitable for purposes of policymaking because it showed the causal structure of the economy and therefore provided points of application for policy. Friedman responded that the fact that the structural model failed when compared to the far simpler naive models meant that the structural model had not, in fact, convincingly isolated the causal structure of the American economy; if the simple model produced better predictions, why rely on a complex, inferior model?

At the same time, Friedman realized that the meaningfulness of the test that had been performed was limited. After all, only the predictions for one or two years had been looked at and it would be rather too hasty to draw far-reaching conclusions purely on that basis. In later years, the Cowles Commission would link the quality of a model to, for example, its ability to show the characteristics of empirical data series, such as the length of business cycles, peaks and troughs and so on. Nonetheless, Friedman's criticism brings us back to the fundamental problem that it is difficult if not impossible to determine whether a structural model has indeed identified the structure of an economy.

Friedman concluded that the development of such large-scale models was profoundly ill-advised. He made clear that this did not mean he was opposed to a mathematical–statistical approach to economics, but such research would, he believed, do better to concentrate on modelling specific, individual markets. It might then be possible to link those studies together in order to throw light on the economy as a whole. There was a long way to go before macro-economic structural models could be of any use or significance.

Friedman's criticism had consequences for the way economic policy is thought about. We have seen that Christ defended econometric structural models by pointing to their usefulness in policymaking. Friedman believed this was precisely the purpose for which econometric models were unsuitable. This tallied with the conviction already dominant at the economics faculty in Chicago that markets were better off without state intervention. In this sense, the Chicago economists were diametrically opposed to the economists and engineers of the Cowles Commission. Whereas the Cowles Commission was hoping to deploy its research actively for purposes of policymaking, the Chicago economists saw the failure of econometric models as a reason to let markets go their own way. Behind Friedman's shrewd and justified technical criticism of Christ's conclusions, based on the test of the Klein model, lay an ideological difference of insight about the role of the state in a market economy. The distinction between these two aspects (technical and ideological) was to play an important part in Friedman's renowned essay 'The Methodology of Positive Economics' of 1953, which we will look at next.

6.6 Milton Friedman's essay 'The Methodology of Positive Economics'

Aside from Mill's essay of 1836, no essay on the methodology of economics has been more quoted and argued over than Friedman's 'The Methodology of Positive Economics', published in 1953. The main reason for this is the immense provocation contained in the article, where Friedman writes that the assumptions on which economic models are based are totally irrelevant and that the only value of such models lies in their ability to predict. Friedman even goes so far as to claim that the further the assumptions are from reality the better. Why this provocation? And why was this a stance that many economists (though far from all) ultimately pointed to with approval?

Earlier in this book, we saw that John Stuart Mill described economics as a science that studies an aspect of human dealings, namely the human being in his quest for prosperity, his preference for the immediate consumption of luxuries and his aversion to work. Mill acknowledged that this was an abstraction: 'Not that any political economist was ever so absurd as to suppose that mankind are really thus constituted, but because this is the mode in which science must necessarily proceed' (Mill, 1967, p. 322). We have also seen that the reason Mill gave for working with an abstraction was the complexity of economic reality. There were so many 'disturbing causes' that economists could make scientific pronouncements only by concentrating on one aspect of human motivation. The controlled experiment, used with such success in the exact sciences, was unfortunately unavailable to the economist. Mill then defended the idea that people actually do act according to this one aspect by calling on the reader to perform an experiment on his or her own motives for acting. He or she would inevitably come to the same conclusion and see that the quest for prosperity was an important motivation. Mill's argument therefore ultimately rests on a reality claim about motives for action that could be tested by introspection. Introspection was an experiment upon an individual mind, from which a general inference could legitimately be drawn.

Friedman explicitly declined to use this argument. In this and the previous chapters, we have seen how the criteria for what counted as empirical evidence shifted towards statistical data as the science of economics developed. So in the twentieth century, if an economist wanted to declare that people in fact allow themselves to be led by their desire for prosperity, he had to provide statistical evidence. He could no longer point to a kind of mental experiment that a person could carry out on himself that would then, as if by magic, gain universal validity. In this connection, I have already mentioned the scornful reference by Francis Ysidro Edgeworth to those 'introspective marks of brain activity' to which he said Mill had appealed. Certainly in the United States after the Second World War, an economist who (like Lionel Robbins) relied upon introspection was bound to meet with derision from the profession.

But Koopmans' criticism of Burns and Mitchell makes clear that even an empirical statistical approach to economics was not without its problems. How would it ever be possible to use statistics to identify the underlying mechanism of an economic system? We have seen that this was a difficulty when it came to assessing structural equations as well. The solution suggested by Koopmans was that, conversely, you must already have some idea of the mechanism in order to be able to assess it statistically. This quest for an underlying mechanism, for invariant behavioural equations, is illustrated in Tinbergen's work.

In his criticism of Burns and Mitchell, Koopmans suggested that economists could use questionnaires to gain an understanding of the behaviour of actors in an economy. It so happened that two economists from the Oxford Research Group, Robert Lowe Hall and Charles J. Hitch (who became director of the RAND corporation after the war), had used lists of questions to investigate the truth of one of the most important assumptions made by economists, namely that a person running a business tries to maximize his profits. They published their decidedly negative results in 1939. Instead of a marginal price strategy (which would have supported the assumption of profit maximization), most businesspeople answered that they used the strategy of cost-plus pricing. Hall and Hitch drew from this the far-reaching conclusion that entrepreneurs were not able to follow a marginal price strategy because at no point would it be clear what such a strategy would mean in practice. Koopmans' casual suggestion that questionnaires could function as an independent source that would confirm the behavioural assumptions of economists was therefore misplaced. If the questionnaires had rendered up information at all, then the only possible conclusion was that the assumptions made by economists about behaviour would have to be revised.

In his essay, Friedman made clear that this conclusion was a step too far for him. He certainly did not claim that the results of questionnaires were worthless. They were 'extremely valuable' in suggesting new hypotheses in cases where predicted and observed behaviour diverged. But however useful in that sense, they were 'entirely useless as a means of *testing* the validity of economic hypotheses'. Friedman argued that the answers given by a businessman (or any other economic actor) in a questionnaire intended to investigate his behaviour did not straightforwardly offer insight into his motives for acting. Why would an economist take at face value the response that he used cost-plus pricing? Why would a businessman tell an economist the truth? For an economist, it had no relevance if a businessman claimed he used a different price strategy from that which was consistent with profit maximization, or indeed made any other claim about his own behaviour.

To Friedman, what was significant was the behaviour shown by a businessman in a situation relevant to an economist, which is to say within the constraints of the market. It does not matter very much what an economic actor says; what matters is his actual behaviour in those circumstances. Friedman claimed there could be little doubt that in a market environment the only businesses that would survive were those that de facto maximized profits, irrespective of what the people running them said. After all, if they acted any differently, the business would fail. In short: the market is always right.

According to Friedman, a successful businessman acts *as if* he is out to maximize profits, whatever he might have to say on the matter and irrespective of whether he is aware of it or not. Whether a theory's assumptions about behaviour were realistic was irrelevant. All that mattered was whether the theory's predictions were more accurate than those of a rival model.

Friedman went on to stress the same criterion for the assessment of the 'correctness' of a theory as he had in his criticism of the test of the econometric model applied by the Cowles Commission. A theory cannot be judged in isolation; it always needs to be compared with another theory or other theories. Friedman offered a number of criteria for use in making such a comparison. How simple is the theory? How 'fruitful' is it, or in other words: to how many different fields can it be applied? But the most important criterion was a theory's predictive power. Which theory makes the best predictions, new or otherwise?[2] The theory that makes the best predictions is the best theory. Friedman also stressed that we can never 'prove' that a theory is 'correct', only that it is worse than a rival theory.

6.7 Friedman and Popper

Because of this last point – theories cannot be proved but only disproved – Friedman's methodology has often been compared to that of the philosopher of science Karl Popper, who was gaining influence in the Anglo-Saxon world in the same period. This book is not the place to look closely at Popper's theories, so the reader is referred to the various excellent general introductions to the philosophy of science that are available. Nonetheless, for anyone attempting to gain a good understanding of Friedman's position, a comparison with Popper is illuminating.

In essence, Popper claimed that scientific theories should be regarded as hypotheses that need to be tested empirically. If a hypothesis withstands such a test, then we can accept it, but only for the time being. After all, a counterexample may yet be found. But a counterexample is only a counterexample if it fulfils the conditions under which a scientist thinks he can find a connection between different variables (the hypothesis). For a test, therefore, it is always necessary to state the conditions under which the hypothesis leads to the predicted outcome. These are known as the initial conditions.

Does it follow from a successful test that the theory is true? The dominant movement within the philosophy of science up until then, logical positivism, gave as its criterion for 'good' science that it must be based on verifiable empirical claims; laws were derived from observations and in turn those laws were tested against empirical data. This circle from empiricism to theory (laws) and back to empiricism guaranteed, at least in the early formulations of logical positivism, that scientific knowledge was certain knowledge. Popper's view of science put an end to this belief in the certainty of scientific knowledge. His surprising objection to it was that all you could say with certainty was that a theory that failed a test had been refuted or, to use Popper's own terminology, falsified.

Although Popper's original book, *Logik der Forschung*, had been published in 1934, it is unlikely that Friedman knew of it. The English translation appeared in 1956, under the title *The Logic of Scientific Discovery*, and only then did it begin its triumphal march through the philosophy of science. It seems more likely that Friedman based his ideas partly on the criticism by Irish economist Terence Hutchison in the 1930s of Lionel Robbins' essay on the method of economic science. This criticism, published in 1938 as *The Significance and Basic Postulates of Economic Theory*, was strongly influenced by logical positivism, but it contained ideas similar to those that Popper was developing at the time. Like Popper, Hutchison stressed the importance of an empirical, statistical test of economic theory. To him, such a test meant that the assumptions behind an economic theory were subjected to testing, such as the testing carried out by Hall and Hitch in their research into the actual price-setting behaviour of business proprietors. Their test had produced a negative result. How could it be that such a result was seen as unimportant by an empirically oriented economist such as Friedman?

To understand this, it is important to look at the frame of reference underlying Popper's (and to a lesser degree Hutchison's) philosophy of science, namely experimental science. An experiment can be described as a controlled intervention in a system. If a theory lies behind that intervention, then it is possible to say that the theory is being tested. This was certainly the image of science that Popper had in mind. But not Friedman.

Whether or not Friedman knew about Popper's ideas is of purely historical importance, but it is clear that in his article Friedman uses experimental science merely as a reference point, claiming that it gives a distorted picture not just of a social science such as economics but of the natural sciences, too, such as physics or biology. Science does not fit the image suggested by the controlled experiment. Friedman gives three examples to clarify this point, one from physics, one from biology and one from psychology. The aim of this tour of the different sciences was to show that economic theories are not representations of reality but instruments that work or do not work. So theories ultimately have to be judged by their predictions.

Friedman claims that in the case of economics, we have no choice but to be agnostic about the assumptions of a theory; an economist does not find himself in the situation of a scientist in a laboratory who can determine the initial conditions and then carry out a controlled experiment. An economist cannot rely on the assertions of actors in the market, whether they be the results of introspection or answers to lengthy and detailed questionnaires. All an economist has available to him are market outcomes, actual results. He is therefore forced to make bold assumptions, and the only criterion that he can then introduce to defend those assumptions is that they render up the best predictions – better than the predictions of an alternative model. This says nothing about the truth of those assumptions. In fact, Friedman tells his readers provocatively, it is actually true to say: the more unrealistic the assumptions the better.

Friedman establishes his three examples, from physics, biology and psychology, and then makes his move towards economics. If we use Newton's law of gravity to calculate how long it will take a stone to fall to the ground, does this mean that we truly assume the stone falls in a vacuum? Or that we have controlled for disturbing factors such as differences in air pressure, wind speed, temperature and so forth? If we assume in calculating the angle of leaves in relation to the sun that they maximize the amount of sunlight they receive, do we then assume that leaves consciously turn themselves to achieve a certain angle? If we use geometry to calculate the angle of a shot that a good billiards player will use, do we then assume that the player himself makes the same calculation? The answer in all these cases, says Friedman, is no. Of course we don't assume the stone falls in a vacuum, but that does not make Newton's law unusable. Of course there must be some kind of mechanism that lies behind the turning of leaves to face the sun, but that does not make our assumption unusable. Of course a good billiards player thinks differently from those studying the angle of his shots, but simple geometry can nevertheless serve as a model for his thinking process. A model does not create pictures of the world, rather it is an instrument we use to discover things about the world, an instrument that 'works'.

The same is true of economic models. No economist will claim that all businesses consciously maximize profits. Nonetheless, they can be seen in very many circumstances as if that is what they are doing, as if they know how supply and demand function, understand price elasticity and so on. This does not mean the assumptions made by the economist are true, which is to say actually used in making decisions, or that they could be directly tested in any way. The science of economics is after all not an experimental science (and, Friedman's examples imply, there is nothing wrong with that). But it does mean that based on these assumptions, we can construct economic models that render up good predictions of market behaviour.

To those engaged in the philosophy of science, Friedman's stress on the 'as if' character of economic models was (and is) an important bone of contention. At various times, philosophers have criticized his provocation about the unrealistic content of assumptions. Assumptions may relate to factors that are negligible in one case but not in another, such as the breath of wind that changes the fall of a feather but not of a brick. Assumptions may relate to the range of applicability of models. An assumption may have a purely heuristic function, in the sense that the researcher knows that an assumption is incorrect but abandons it only in a later phase of his research. Such criticisms clarify the character of the assumptions on which economists base their building of models. But in essence this kind of criticism fails to take account of the radical nature of Friedman's provocation, since it assumes that economists are searching for a satisfactory representation of their object of study, an idea reinforced by the notion that economics conforms to the image of a laboratory science. If an experiment is performed in laboratory conditions under controlled circumstances and the outcome is as predicted, then, for example, there are reasonable grounds for assuming that a Z boson exists, or that we have unravelled the structure of DNA. Such was the image of science from logical positivism up to and including Popper. It was also the image of science the Cowles Commission had in mind, despite an awareness of its problematic character. Carl Christ wrote:

> When one thinks of science, one usually thinks also of experiments. In a typical experiment, there is one variable whose behavior is studied under various conditions. The experimenter fixes at will the values of all the other variables he thinks are important, and observes the one

in which he is interested. He then repeats the process, fixing different values of the other variables each time ... The experimenter hopes to find a single equation that describes closely the relationship he has observed.

(Christ, 1951, pp. 35–6)

The Cowles Commission wanted to see its econometric structural models in this way too, but here the situation was more complex:

In more complicated situations there may be more than just one relationship among the variables studied ... there [may be] two or more variables that are not fixed in advance; there may even be no experimenter. Economics abounds with such situations. The simplest is of course a competitive market, in which neither price nor quantity is fixed in advance. The economist assumes that two relations between these variables, a supply equation and a demand equation, must be simultaneously satisfied. The econometric work discussed here is based on the belief that we will do well to make our theory conform to this state of affairs.

(Christ, 1951, p. 36)

This image of economics as going in search of a surrogate for an experiment was not shared by Friedman. Economics was not an experimental science. Economists were not able to establish boundary conditions for a controlled experiment. As a result, the aim of economic models was not to create representations of reality. Models were instruments for analysing the world, but they did not offer representations of it. Friedman pointed with approval to the work of the British economist Alfred Marshall (who taught Keynes at Cambridge), for whom mathematics was an 'engine', an instrument for analysing the world, not a camera. The problem is that a model attempts to be both, not just an instrument but a representation as well. It is a problem that will remain with us throughout the rest of this book.

6.8 Economics between empiricism and ideology

Anyone who examines it closely will notice that in his essay Friedman still opts for the research programme of his teachers Burns and Mitchell. In a footnote, he thanks Arthur Burns and his colleague at the Chicago economics faculty, George Stigler, for their comments, and indeed both the empirical slant of the NBER and the ideological hand of the Chicago School are clearly visible in Friedman's work. Burns and Mitchell had understood, better than the Cowles Commission itself, that the type of structural models Cowles had developed overestimated the feasibility of controlled empirical research. If a comparison with two simple, naive models showed that they worked better than a complex model, did that not in fact show that the *natural selection* [handwritten margin note]

Cowles Commission was wrong to think that it, as a kind of experimenter, had isolated the causal structure of economic reality and could go on to manipulate it for policymaking purposes? The comparative test demonstrated quite clearly that this was not the case. As far as Friedman was concerned, Mitchell was right: an economist must use theories opportunistically. Instead of striving for an all-embracing analysis of an economy, the economist would do better to concentrate on analysing actual markets.

The snag lies in Friedman's emphasis on the accuracy of predictions produced by economic theories as a criterion for choosing which theory to adopt. The financial crisis of 2008 reminded us that economic theories have a dreadful reputation as far as their predictive powers are concerned. This has caused great merriment among scientists in particular. There is of course no dispute about the contrast between a predicted solar eclipse by which you could set your watch and an economic prediction that may be out by many percentage points, but the problem with Friedman's stress on the predictive powers of economic theories goes deeper, as can be seen from his example of a businessman who seems to display profit-maximizing behaviour in order to survive in the market.

This example demonstrates that although Friedman continually emphasized the predictive value of models and the importance of empirical statistical research, he believed that statistical evidence was ultimately unnecessary. Reliance on the assumption of profit maximization was based, in Friedman's own words, on 'evidence of a very different character'. Is it conceivable, after all, that a businessman who did not maximize his profits would survive in the market? To Friedman, that was 'unlikely'. His business would have been eliminated a long time ago by 'natural selection'. Whatever someone running a business said, however much the assumptions in economic models concerning competition might contradict reality, markets compelled profit maximization. Would it be possible for a business to survive if the person running it did not maximize its profits? No. So that was what he did.

That an empirical test for economic theories is actually irrelevant was clear from Friedman's third example, the good billiards player who takes a shot *as if* with the help of geometry. Friedman borrowed this example from an article he had written along with Jimmie Savage, a mathematician and the originator of the way probability theory is used in statistics today. They compared the billiards player with someone who makes rational choices. Such a person will act as if he is applying the logical rules of choice theory. It is important to note that in their article they were referring not to any old billiards player but to an expert, so they were formulating a normative theory, a theory that shows what someone ideally needs to do to make the most rational choice. Whether an actual person complies with this norm is a different matter, and it is only at that point that empirical research begins. Because Friedman's article lays such stress on positive, that is to say empirical, science, the reader might easily fail to notice that his examples shift from positive to normative theory. When he then goes on to introduce the profit-maximizing businessman as an example of the way in which economists construct their theories, he is talking about a normative ideal, in which in the end the market is always right.

Friedman's criticism of Cowles and his defence of economics as an 'as if' science therefore ultimately rest on an ideological choice. Like his colleague George Stigler, Friedman was convinced that free markets were better able to coordinate the actions of individuals than markets in which the state intervened. No great macro-models or, in later years, Walrasian models of general economic equilibrium were needed to show this. After all, how else could a business survive? A *reductio ad absurdum* was enough.

6.9 In conclusion

Milton Friedman's 'The Methodology of Positive Economics' became the most cited methodological essay about economics of the post-war period. For widely divergent reasons, it was an ideal reference point for philosophers of science, economic methodologists and economists themselves.

Philosophers of science recognized the strong relationship between the scientific assumptions of Friedman and those of Popper (and of logical positivism). They concentrated on the problems surrounding the infamous *ceteris paribus* clause ('all other things being equal') that was so important for a controlled test, on the degree of realism of scientific theories and on the question of how to discriminate meaningfully between different theories.

Friedman had provided economic methodologists with a serious essay that enabled them to show that economists were not blind to the many questions and problems surrounding empirical economic research – indeed, from their side too the relationship between Friedman's vision and that of Popper was stressed, and Popper's scientific model was then used as a way of criticizing the actual research practices of economists, who were said to pay too little attention in their empirical research to the obligation to actually test their theories.

Economists, finally, were reassured by Friedman's article that the often heard criticism that they used unrealistic assumptions was groundless. After all, economics was not an experimental science. Merton Miller, an important theoretician of financial markets at the Chicago School, once said in an interview that no one there was worried about the assumptions on which they based their work: 'We just sort of take [Friedman's viewpoint] for granted. Of course you don't worry about the assumptions.' What was important was not how realistic the assumptions were, but the predictive power of the models.

In contrast to practising economists, philosophers of science and methodologists of economics started out from the image of experimental science as an ideal basis for comparison with economics. Where the research practices of economists failed to match this image, the philosophers were quick to conclude that economics was using a garbled version of the 'correct' scientific method. They went on to use a reinterpretation of Mill's essay as a way of defending the econometric work of the Cowles Commission. In its structural models, the Commission was rightly attempting to look 'under the hood', whereas the implication of Friedman's view seemed to be that it didn't really matter what the engine looked like, just as long as it worked. But this criticism of Friedman assumes that it must be possible in principle to see the mechanism that underlies economics in the same way as an experimental scientist is able to isolate the mechanism that, for example, leads to cell division.

It was precisely this reasoning that was radically undermined by Friedman's criticism of the Cowles Commission. Economists were left with simply a test of outcomes, and an ideological conviction that went practically unchallenged, especially in the United States, namely that markets do their work more efficiently than any government could, because economic actors are free to make their choices and appraisals on their own behalf.

Notes

1 Koopmans agreed with John Stuart Mill's interpretation of Kepler's accomplishment, an interpretation that was strongly criticized at the time by William Whewell. The most astute criticism of the idea that Kepler was merely 'summarizing' Brahe's observations can be found in Norwood Russell Hanson's magnificent *Patterns of Discovery: An Inquiry into the Conceptual Foundations of Science*, published in 1958, a book that was eclipsed by Thomas Kuhn's *Structure of Scientific Revolutions* (1962).

2 A new prediction might relate to situations from the past that are unfamiliar to us (such as the existence of a common ancestor for man and apes, or a drop in productivity prior to the seventeenth-century tulip crisis). Think for example of Friedman's proposal that Tinbergen's business-cycle model should be tested against a different collection of data from that used to construct his model.

References

Christ, Carl F. (1951). A Test of an Econometric Model for the United States, 1921–1947. *Conference on Business Cycles*. (New York: National Bureau of Economic Research).

Mill, John Stuart (1967 [1836]). On the Definition of Political Economy; and on the Method of Investigation Proper to it, in *Collected Works of John Stuart Mill*, J.M. Robson (ed.), vol. 4. (Toronto: University of Toronto Press).

7 Modelling between fact and fiction
Thought experiments in economics

7.1 Introduction

In Chapter 6, we saw that Friedman's 'as if' methodology was embraced by economists for two reasons. First, Friedman relieved economists of any worries about the reality content of the assumptions on which their theories were based. His concept of research confirmed what economists had regarded as characteristic of their discipline ever since Mill: economics was not an experimental science, so economists needed to use other criteria for constructing theories and testing them. But whereas Mill relied on a kind of inner experiment in testing the validity of the assumptions underlying his theories, Friedman showed that the validity of assumptions was irrelevant. Economics was not alone in relying on assumptions that everyone knew from the start to be unrealistic; disciplines such as astronomy and biology did the same. This did not make their theories any less workable. The usefulness of economic theories could be measured by their power to make accurate predictions.

The second reason why economists, especially in the United States, embraced Friedman's methodology was that it made faith in the efficient working of markets, which was widely shared among economists, into a basic principle of good economic science. We have seen that this second criterion was not empirical but normative, and it was therefore in fact at odds with the stress that Friedman laid on predictions rather than assumptions. That the market worked was not disputed and did not need to be tested. This was the firm belief of economists of the Chicago School, who in the 1990s (the decade of neoliberalism) carried off one Nobel Prize for Economics after the other.

But the Cowles Commission and economists with Keynesian or social democratic leanings remained interested in the question of whether, and if so how, the market worked. Some of these economists became lost in complex mathematical proofs of what was known as general equilibrium theory, a programme of research that attempted to show under which conditions a complex system of markets will reach an overall equilibrium. In the early 1980s, it found itself at a dead end. Others, often taking a

Keynesian approach, used simple mathematical models to ask fundamental questions about the workings of economies.

Paul Samuelson was one of them. Along with Friedman, Samuelson is undoubtedly one of the most important post-war American economists, but he was more a Keynesian interventionist than a believer in the free market; indeed in the early 1960s he served as an advisor to President John Kennedy on economic affairs. Samuelson believed the market did not always work, and that it was the task of economists to consider the conditions under which markets failed. Samuelson cherished the image of the economist as a theoretician that we encountered in Chapter 3 in looking at Lionel Robbins. But whereas for Robbins theory was still mainly a matter of words, Samuelson took a quite different view. His Harvard PhD dissertation 'Foundations of Analytical Economics: The Observational Significance of Economic Theory', which was the basis for his 1947 book *Foundations of Economic Analysis*, helped to ensure that henceforth the 'natural language' of economists was mathematics. This did not mean that to Samuelson economic theory was merely a game of tautologies. On the contrary, as will be illustrated in this chapter, many of Samuelson's contributions to the science of economics can be regarded as thought experiments that teach us something fundamental about the world in which we live. This cuts across a distinction carefully nurtured today between mathematics and logic on the one hand, as analytical instruments with which only tautologies can be formulated, and on the other the empirical research we rely on to breathe life into them. Samuelson clearly had a different approach. His mathematical thought experiments, like many other thought experiments, revolve around a paradox, the solution to which teaches us something about the world.

7.2 Paul Samuelson's 'Exact Consumption-Loan Model'

As an example I have chosen 'An Exact Consumption-Loan Model with or without the Social Contrivance of Money', which Paul Samuelson published in the *Journal of Political Economy* in 1958. I have chosen this article because it was thoroughly scrutinized by philosopher of science Daniel Hausman in 1992, and his discussion will later serve as a reference point in my efforts to clarify the difference between thought experiments and the testing of empirical models. Between the wars, Paul Samuelson studied at Harvard, where his teachers included Joseph Schumpeter and Wassili Leontief. The story goes that after Samuelson finished his oral exam in his first year of postgraduate study, Schumpeter asked the other members of the committee, 'Well gentlemen, did we pass?' Samuelson defended his dissertation and in 1941, after some delay caused by the Second World War, published it in 1947 as *Foundations of Economic*

Analysis. The book marked a definitive end to the literary, discursive approach to economics. From this point on, mathematics was the language in which the discipline preferred to express itself.

Samuelson was an active participant in the many interdisciplinary study groups at Harvard and he was especially taken with the 'operationalism' of physicist Percy Bridgman, which involved defining scientific terms according to the operations they measured.[1] A unit of measurement such as a metre is defined not according to its essence but according to convention. Inflation is defined by the rules for measuring inflation, no more and no less. In practice, Samuelson seemingly paid little attention to empirical research, portraying himself as a theoretician who made purely abstract mathematical models (an image that is rather remarkable, incidentally, given his extensive involvement in empirical research projects from the 1940s until well into the 1960s). Did this mean that such models had no empirical implications? Let us look more closely at a page from Samuelson's 'Exact Consumption-Loan Model'.

The page is taken from an article in which Samuelson constructs an extreme economy. In the world of Samuelson's model, there are no durable goods and the lifetime of workers is divided into three phases, two in which they work and one in which they are in retirement. All goods that are produced endure for one period only and therefore cannot be saved up for the next period. There is no medium of exchange (money). Since there are no durable goods, the only items that can be exchanged are those produced in the same period. This model is referred to nowadays as Samuelson's model of overlapping generations (at any given moment, three different generations are alive, at least as long as there is no beginning or end to the chain as a whole). The idea of overlapping generations is often used by present-day economists to model an economy, so it has undoubtedly found acceptance in economics as a profession, but does Samuelson's model say anything about empirical reality?

The question that Samuelson investigates in his article is whether he can find in his very strange economy a pattern of individual savings (and the interest rates that go with them) that can be brought into being purely by market transactions and with which everyone is content. The surprising answer is: no. That answer was not just a surprise to Samuelson, it ought to amaze any economist. After all, it is a central tenet of economics that market forces are beneficial, and among economists it is usual (although historically incorrect) to reduce this doctrine to Adam Smith's metaphor of an 'invisible hand' that guides people's behaviour in the market such that the outcome is optimal for everyone.

though perhaps it falls short of explaining the remarkable quantitative identity between the growth rates of interest and of population.

THE INFINITY PARADOX REVEALED

But will the explanation survive rigorous scrutiny? Is it true, in a growing or in a stationary population, that twenty-year-olds are, in fact, overconsuming so that the middle-aged can provide for their retirement? Specifically, in the stationary case where $R = 1$, is it necessarily true that $S_1(1, 1) < 0$? Study of $U(C_1, C_2, C_3)$ shows how doubtful such a general result would be; thus, if there is no systematic subjective time preference so that U is a function *symmetric* in its arguments, it would be easy to show that $C_1 = C_2 = C_3 = \frac{2}{3}$, with $S_1(1, 1) = S_2(1, 1) = +\frac{1}{3}$ and $S_3(1, 1) = -\frac{2}{3}$. Contrary to our scenario, the middle-aged are *not* turning over to the young what the young will later make good to them in retirement support.

THE TWO-PERIOD CASE

The paradox is delineated more clearly if we suppose but two equal periods of life—work and retirement. Now it becomes *impossible* for *any* worker to find a worker younger than himself to be bribed to support him in old age. Whatever the trend of births, there is but one equilibrium saving pattern possible: during working years, consumption equals product and saving is zero; the same during the brutish years of retirement. What equilibrium interest rate, or R, will prevail? Since no transactions take place, $R = 0/0$, so to speak, and appears rather indeterminate—and rather academic. However, if men desperately want *some* consumption at *all* times, only $R = \infty$ can be regarded as the (virtual) equilibrium rate, with interest equal to -100 per cent per period.[7]

We think we know the right answer just given in the two-period case. Let us test our previous mathematical methods. Now our equations are much as before and can be summarized by:

Maximize $U(C_1, C_2) = U(1 - S_1, 0 - S_2)$

subject to $S_1 + R_t S_2 = 0$.

The resulting saving functions, $S_1(R_t)$ and $S_2(R_t)$, are subject to the budget identity,

$$S_1(R_t) + R_t S_2(R_t) \equiv 0 \text{ for all } R_t . \quad (4')$$

Clearing the market requires

$$0 = B_t S_1(R_t) + B_{t-1} S_2(R_{t-1}) \quad \text{for} \quad (5')$$
$$t = 0, \pm 1, \pm 2, \ldots .$$

If $B_t = B(1 + m)^t$ and $R_t = R_{t+1} = \ldots = R$, our final equation becomes

$$0 = B\left[S_1(R) + \frac{1}{1+m} S_2(R) \right]. \quad (8')$$

The budget equation $(4')$ assures us that equation $(8')$ has a solution:

$$R = \frac{1}{1+m} \quad \text{or} \quad m = i .$$

with $\quad 0 < S_1(R) = -RS_2(R) .$

So the two-period mathematics appears to give us the same answer as before—a biological rate of interest equal to the rate of population growth.

Yet we earlier deduced that *there can be no voluntary saving in a two-period world*. Instead of $S_1 > 0$, we must have $S_1 = 0 = S_2$ with $R = +\infty$. How can we reconcile this with the mathematics?

---

[7] A later numerical example, where $U = \log C_1 + \log C_2 + \log C_3$, shows that cases can arise where no positive R, however large, will clear the market. I adopt the harmless convention of setting $R = \infty$ in every case, even if the limit as $R \to \infty$ does not wipe out the discrepancy between supply and demand.

Figure 7.1 Fragment of text from Paul Samuelson, An Exact Consumption-Loan Model with or without the Social Contrivance of Money.

Source: *Journal of Political Economy*, 66(6) (1958), p. 474.

With this in the back of our minds, let us look in more detail at the page from Samuelson's article shown in Figure 7.1. Every economist will recognize from the equations the standard micro-economic theory that Samuelson uses. In his own book *Foundations*, he had made the maximization of utility on a budget restriction part of the standard equipment of an economist. We also see the equation in which all markets are cleared, in other words in which all supply and demand orders are filled and the savings plans of all the actors in Samuelson's economy can be fulfilled. The solution renders up an interest rate that is equal to the growth rate of the population. Samuelson discusses the shape of preference orderings and assumptions about the time preferences of the actors in the economy. All this was part of the standard equipment of the academic economist in the 1950s, as it still is today.

Scientists and economists outside of the mainstream look upon this practice with suspicion. Herbert Simon, who later won a Nobel Prize for Economics, complained as early as 1954 that the science of economics made all kinds of psychological assumptions from the 'comfort of the armchair'. In that same year, one of the founders of cognitive psychology, Ward Edwards, wrote that economists generally used 'an armchair method' in their efforts to understand human behaviour: 'They make assumptions, and from these assumptions they deduce theorems which presumably can be tested, though it seems unlikely that the testing will ever occur' (Edwards, 1954).

It does indeed seem improbable that Samuelson would ever actually have wished to test his mathematical model, not only because of the assumptions he made but because he showed that the workings of the market mechanisms in his model economy resulted in a paradox. There is a market equilibrium, but no means can be found by which it is possible to achieve this market equilibrium through market transactions: the 'invisible hand' does not work.

The paradox becomes all the more clear if the life of each of the actors in the model economy consists not of three but of just two periods. An actor is either working or retired. If he works, he produces a product but that product can be consumed only in that same period (after all, there are no durable goods). If he has retired, he produces nothing at all. How on earth could a retired person persuade someone who was working to hand over part of the product he was working to produce? Anyone who, like an economist, is sceptical about the altruistic motives of human beings and tends to believe that people are driven by self-interest, could only answer: nothing at all. It was therefore *'impossible* for *any* worker to find a worker younger than himself to be bribed to support him in old age' (Samuelson, 1958). The only theoretically possible interest rate Samuelson could think of that would mean everyone at least had something to consume was a negative interest rate of minus 100 per cent per period. Samuelson realized that such an interest rate was purely academic.

Since this interest rate was no more than a matter of theoretical curiosity, it is not hard to imagine why philosopher of science Daniel Hausman concluded in his in-depth review of Samuelson's model that this was a prime example of the way in which economists had closed themselves off from the 'real world' very soon after the Second World War. They had given priority to elegant scientific models over 'the hard work of other social scientists who have to seek sometimes almost blindly for significant causal factors' (Hausman, 1992, pp. 118–19). Samuelson's model did indeed seem to be an example of a study that was neither based on meticulous research in the field nor amenable to the instruments of the econometrician and therefore to statistical testing. The model seemed utterly devoid of empirical content of any kind. It was not just 'armchair economics' because of the suppositions it relied upon but also, and perhaps primarily, because of the idiotic paradoxes with which this model world became entangled. It was not at all clear how such paradoxes could ever teach us anything about the real world. It seemed a purely analytical, formalistic mental exercise, detached from reality.

Was it in fact true that, very soon after the Second World War, mathematical economists found themselves in the equivalent of a photographer's darkroom in which they told only mathematical truths that had nothing to do with the real world? Or does Samuelson's model tell us something about the world after all, something that we perhaps too easily overlook in our daily lives but that is nevertheless fundamental to any understanding of the workings of a market economy? The mere fact that Samuelson's model is still used by economists today should give us pause.

7.3 Virtual journeys and thought experiments

Let me take a step back and look again at our example from Chapter 2, Robert Malthus' *An Essay on the Principle of Population*, to get a firmer grip on these questions. In his essay, Malthus was responding to the utopian order described by political thinker William Godwin in *An Enquiry Concerning Political Justice* of 1793. Godwin wrote about his utopia at a crucial historical juncture. After the French Revolution (and the subsequent period known as the Terror), not just France but all of Europe was in turmoil. The general public needed a hopeful message, which is what Godwin offered in his book.[2]

Godwin wrote that the source of all evil lay in social institutions, but not in human nature. In this starting point, an echo can be heard of Jean-Jacques Rousseau and the Enlightenment. Godwin's utopia, as we have already seen in broad outline, was a society not based on private property. The institution of marriage would be abolished. Altruism would prevail over self-interest and the problem of overpopulation, to which Godwin was far from blind, would be resolved of its own accord, because of a lessening of the 'commerce between the sexes'. Godwin even went so far as to predict that individual life expectancy would increase to infinity as a result of all these changes.

Malthus' strategy was to assume that Godwin's utopia had been realized in full. He then showed that the population would very probably increase more quickly than food production. This would lead to a fight over food, and the institutions Godwin detested would quickly be reinstated.

Malthus' reasoning is an example of a thought experiment, since it presumes a specific state of affairs. In your mind you change one or more factors and then declare what you believe the consequences of that change or changes would be. In many cases, a thought experiment concerns a paradox. It is not a real experiment because you do not actually manipulate factors, you do so only virtually, or mentally. This may be, although not necessarily, because it is impossible to exclude or control certain factors in reality.

The strength of Malthus' thought experiment was that he fundamentally disrupted a line of argument that had seemed plausible (with the exception of eternal life) to adherents of Rousseau and the Enlightenment if no one else. He showed that it led to a fundamental paradox that undermined the basic assumptions on which Godwin's utopia rested. A society without institutions, where the governing principle of everyone's dealings was altruism, could not exist. It was purely a product of the imagination: 'We have supposed Mr. Godwin's system of society once completely established. But it is supposing an impossibility. The same causes in nature which would destroy it so rapidly, were it once established, would prevent the possibility of its establishment' (Malthus, 1986, p. 210).

Whereas Godwin had been able to attribute the population problem to the imperfection of social institutions, in Malthus' hands it became an inescapable law of nature.

Thought experiments were used (as indeed they still are) not only to discover the principles behind society but to contemplate the laws of nature. Most thought experiments have their origins in the natural sciences. One famous example is the following thought experiment by Simon Stevin, engineer and personal advisor to William of Orange, which he used to prove that there could never be such a thing as perpetual motion. He accompanied his experiment with an engraving (see Figure 7.2). It shows an inclined plane with a necklace of evenly spaced beads draped loosely over it. On the longer slope lie four beads, on the shorter slope two. The part of the string of beads that hangs freely below is symmetrical, with four beads on either side.

Stevin's argument ran as follows. If the symmetrical part of the string of beads is cut away, will the rest of it move? You might think it would, since four beads weigh twice as much as two, so the necklace should roll towards the left. In that case, it ought to have moved when the loosely hanging beads were still part of it, but then they were all at rest. So it will remain at rest now as well; the forces cancel each other out. This means that a system to which no movement is added cannot move of its own accord. So a *perpetuum mobile* does not exist.

Figure 7.2 Front cover of Simon Stevin, *Beghinselen der weeghconst* (1586).

It is important to realize that although we might be able to carry out Simon Stevin's thought experiment in the real world (an issue still contested among philosophers to this day), given his argument there is really no point in doing so. Which is not to say there is nothing to be learned from it. On the contrary, it teaches us to think about the character of forces and the rules they obey. Is that purely formal knowledge? No. The aim of a thought experiment is not experimental in the sense that it involves carrying something out in reality, with actual materials, but it does contain empirical knowledge – in other words, it presents knowledge of which we may perhaps be unaware, made explicit by the thought experiment. Has anyone ever seen a loosely draped string of beads move without any force being applied to it?

Stevin was so satisfied with his thought experiment that he added to the drawing the line 'Wonder is no Wonder'. This expresses precisely the paradoxical character of a thought experiment. We can follow the unambiguous logic, but the conclusion nevertheless goes against our intuition. (We intuitively believe that because four beads are heavier than two, the string of beads must move, although that is not the case.) Philosopher of science Thomas Kuhn believed that its paradoxical, dislocating character was typical of the thought experiment, which compels us to see our familiar world in a new light, and this was also precisely what Malthus had in mind in dismissing Godwin's utopia. Intuitively, it might be plausible to attribute social disorder to human institutions, but on further examination the argument did not stand up. What may have seemed trivial or irrelevant to our intuition was in fact of far greater importance – population growth is a law of nature independent of social institutions.

7.4 Shaking up an axiom

In a similar way, physician Bernard Mandeville, a Dutchman by birth, shook the British establishment to its core in the early eighteenth century with his famous *The Fable of the Bees*. Its subtitle, *Private Vices, Publick Benefits*, expressed a startling paradox: even the most selfish behaviour could be in the public interest. Almost all prominent moral philosophers of his day, from Berkeley and Hutcheson, via Hume and Smith to Kant, felt obliged to come to terms with Mandeville's provocation in one way or another. In one of the basic texts of political economy, *The Wealth of Nations* of 1776 (pp. 26–7), we can read Smith's solution, which replaced the early eighteenth-century axiom that human behaviour was virtuous because it was by nature directed towards the common good:

> It is not from the benevolence of the butcher, the brewer, or the baker, that we expect our dinner, but from their regard to their own interest. We address ourselves, not to their humanity but to their self-love, and never talk to them of our own necessities but of their advantages.

In that same book, Smith introduces his important theory of the division of labour by pointing to the paradox that an 'industrious and frugal peasant' is better off than 'many an African king'.

This last example shows that a paradox does not necessarily arise from a carefully constructed thought experiment. It may emerge when the observations and experiences presented in travel accounts (or derived from one's own travels) are compared to the situation at home. Up to a certain point, a thought experiment can be compared to a trip to a virtual world that teaches us something about the world with which we thought we were so familiar. Even in his own day, British political economist Richard Whately, whom we encountered in Chapter 2, believed that the paradox was a vehicle for conceptual renewal in the sciences and that his own discipline in particular was riddled with paradoxes. For Whately, the true political economist was a theorist, looking over the heads of the players and seeking out and explaining the paradoxes in their behaviour. With his *Fable of the Bees*, Mandeville was to Whately a prime example of a person who had shown how self-interest in a market-based society happened to serve the general interest. In Adam Smith's image of the 'invisible hand', this insight gained the status of an unshakeable axiom.

If we turn to Samuelson's extremely successful textbook *Economics*, published in 1948, we will see that he used an everyday example to illustrate this axiom. *Economics* had been written for students of technical sciences at the Massachusetts Institute of Technology (MIT), but it quickly grew to become one of the most popular and frequently used economics textbooks of all time. At the start of the third chapter, Samuelson discusses a 'famous example' (which, without citing any sources, he transfers from London to New York; it had originally been used by Richard Whately). He asks the reader to imagine that goods in New York need to be distributed by some kind of central planner. Such a thing would surely be inconceivable. To Samuelson, this alone was sufficient proof that distribution is better done by individuals acting as individuals, motivated by their own self-interest. Markets coordinate selfish decisions by means of an invisible hand, so that they serve the general interest. But what Samuelson presents in his textbook as a foregone conclusion – the conviction broadly shared among economists that markets work efficiently – was the subject of research in his academic work. Let us go over his model one more time.

7.5 The practical implications of a paradox

At the start of his article, Samuelson presents his subject as purely theoretical. If theory is defined as having a mathematical structure and lacking statistical data, then this is indeed the case. But the subject treated by Samuelson has to do with an issue that is not merely interesting from a theoretical point of view but ought to interest any economist who has faith in the workings of the market, and indeed a broader public. He deals with a very practical question: why would a person want to contribute to a pension that is paid out to someone he or she does not know? What does that person have to gain? This is a real and fundamental issue, as we know from present-day discussions about the problem of an aging population and the viability of the pension system.

In earlier times, Samuelson writes, it was customary for children to look after their parents, but 'that is now out of fashion'. Economists believe that markets can instead optimally coordinate behaviour. Is this true? As a 'test', Samuelson constructs a model world in which 'cold and selfish markets' by themselves have to care for the elderly. No family ties exist between the actors in his model.[3] There is no money and no possibility of savings. What does the market do? Samuelson investigates his model world for a static and for a growing population, and reaches surprising conclusions. Indeed they surprise even him. The market does nothing at all.

It turns out that in theory a social optimum exists in which individuals maximize their savings over their lifetimes, based on a reasonably normal interest rate. But when Samuelson then looks at how such a social optimum is to be achieved, it turns out that his model economy ends up in a paradoxical situation in which although there is a social optimum, it cannot be achieved by market dealings – there is no invisible hand. Economically relevant interest rates (that is, interest rates that enable individuals to get what they need through market trading) diverge considerably from the optimum. There are negative interest rates and even interest rates of minus 100 per cent. (Who would put money into an investment account at that rate of interest? Hands up please!) So Samuelson is forced to conclude that in his fictional world, 'the social optimum configuration can never be reached by the competitive market, or even be approached in ever so long a time'.

Samuelson's fictional model world demonstrates that we cannot rely entirely on markets for a pension system. This undermines the 'fundamental intuitions' of economists about the beneficial workings of markets. Samuelson thereby clarifies why it is that in *our* world there are compulsory pension schemes and, something he looks at in more detail in his concluding section, why such a thing as money is important in an economy. Money can be a means of transferring wealth to the future. As soon as money exists, a market can do its beneficial work. Samuelson's fictional world teaches us lessons about the importance of enforced cooperation and of a stable currency. Such conclusions are obviously not just of importance to the theoretician; they have a real-world political charge. If selfish markets are not going to take care of the elderly, then society needs to ensure there are institutions that will do so.

Economists (and not only economists) like to believe that markets work well, which is to say that they deliver the best outcomes for everyone. Economists and politicians regularly appeal to their own 'experience' to substantiate this belief. But faith and experience will not get an economist or anyone else anywhere in Samuelson's model world. If in that world it is left to the market to achieve an optimal situation, then individuals will be disappointed. The hope that there is an 'invisible hand' able and willing to steer an individual's decisions in the right direction is dashed by Samuelson, and an arithmetical example makes this abundantly clear. We cannot, is the message, simply trust to our own good sense. More specifically, we cannot simply rely on the market to do its job in all circumstances. Without adequate institutions, markets do not work. So it will be no surprise to discover that Samuelson not only advised the Kennedy administration on economic policy but was an outspoken opponent of perhaps the greatest free-market ideologist in the United States, Chicago economist Milton Friedman.

7.6 Conceptual and empirical exploration

At the end of his article, Samuelson compares his working method with the performance of an experiment. In this connection, economists use the expression *ceteris paribus*: if all other conditions remain the same, then the effect of a change to one variable in a model can be determined unambiguously, as it would be in an experiment. The implicit assumption is that the model world and the real world resemble each other in some way, or at any rate exhibit a sufficient degree of resemblance for us to draw conclusions about our world. This is how many contemporary philosophers of science think about models. In the laboratory, attempts are made to isolate a connection between cause and effect, while controlling for other factors. In their models economists isolate a single important aspect of the truth and their model worlds thereby become a fictional analogue for an experiment that could actually be carried out in a laboratory.

In his important book *The Inexact and Separate Science of Economics* (1992), American philosopher of science Daniel Hausman argues that theoretical models such as those used by Samuelson have no empirical pretensions, indeed cannot have any, since use is made of concepts that resist all empirical testing. In Hausman's opinion, Samuelson constructs an artificial, mathematical world and draws conclusions from it. Those conclusions may be surprising or unsurprising, but they relate only to the internal workings of the model. What Samuelson does is therefore purely a matter of conceptual exploration, of research into the conceptual implications of a formal model world. Whether or not the outcomes produced by such a model can be tested empirically (and thereby gain empirical significance) is a separate issue. Samuelson makes no attempt to do so, in fact Hausman believes it is difficult to see how his model could ever be subjected to an empirical test, since the assumptions Samuelson makes are too unrealistic and the outcomes of his models too remote from what we know empirically about our economies. Samuelson's exercise is therefore of purely hypothetical significance, not in the sense of an empirical hypothesis but in the sense that it is a conceptual exploration of an artificial, synthetic world.

This, Hausman claims, is a deficiency of all such models, and he concludes that economists are more interested in the 'elegant finesses' of their mathematical formulations than in the 'hard work' that social scientists carry out in other disciplines. This lack of interest in empirical research finds expression in the way economists focus on the question of whether markets work or not. To anyone who agrees with Hausman, the comparison Samuelson makes between what he does and an experiment can only be misleading.

By calling his book *The Inexact and Separate Science of Economics*, Hausman is referring directly to the work of Mill, who drew a comparison with laboratory experiments. Political economy was not suited to such a method, since you could never control for all the disturbing factors at work in an economy. It was therefore an 'inexact science' that could deduce only 'tendencies' and not, like the natural sciences, numerically precise laws.

7.7 Paradoxes change the world

So is Hausman's verdict on Samuelson justified? At the start of his article, Samuelson makes a number of observations about our world (for example: children in the West no longer look after their parents, so strangers have to be 'bribed' to do so), but at the same time he makes clear that he needs to introduce several rigorous abstractions to make the problem he wants to address visible. Hausman's misconception is that Samuelson is developing a hypothesis that can then be tested empirically.

Nowhere does Samuelson claim that his model represents our world in such a way that it can be tested, either experimentally or statistically. On the contrary, his model world is fundamentally unlike our own. It is precisely by taking us with him into a quite different world that he can offer us insights into the characteristic features of *our* world, features that we, as economists who believe in the beneficial workings of markets, might easily overlook: markets have to be tamed by institutions. This becomes clear in Samuelson's article only when he demonstrates that our 'common sense' leads us to conclusions that cannot be valid (despite all our faith in self-regulating markets, whether or not based on experience). What Samuelson puts to the test is economists' intuition that selfish markets produce efficient results. This is what the model is intended to demonstrate. Only then are we confronted with a surprise: they don't. Markets do not automatically work. The lessons of such models, to quote Whately, 'may be sense but at least they are not *common*-sense' (*Introductory Lectures on Political Economy*, 1832, p. 65).

For this reason, it is more to the point to compare Samuelson's theoretical modelling with Kuhn's thought experiments than with actual experiments. Thought experiments can be seen as a form of conceptual exploration, but they are more than that; they change the way we look at the world, just like the famous example of the old woman and the young lady (see Figure 7.3). It is not so much a matter of a conceptual apparatus that is then refined but rather of a revision of the way we understand the world. A thought experiment begins with everyday observations, then takes us into a world in which those observations are put to the test before returning us to the everyday world armed with concepts that alter the way we see it.

Kuhn stresses that thought experiments are not primarily useful in putting confused or inconsistent concepts to the test (that would merely amount to a clarification of our conceptual apparatus) but in making clear that our observations do not meet our expectations and that those observations must therefore be understood differently (and our expectations of

112 *Modelling between fact and fiction*

Figure 7.3 What do we see, an old woman or a young woman? This picture is a standard example of what is known to psychologists as a 'gestalt switch'; a change in perception changes the world. Clearly it is far from straightforward to distinguish between concept and fact.

them adjusted). This is the reason why Kuhn makes a connection between his thought experiments and his ideas about scientific revolutions. It is precisely because paradoxes have such a central place in thought experiments that it may be a long time before readers are prepared to accept the lessons that need to be drawn.

This applies for example to Mandeville's *The Fable of the Bees*, a book condemned as a 'public nuisance' a few years after it appeared. The assertion that acting out of self-interest could benefit the community as a whole was still anathema in 1714. The fact that this same assertion became one of the basic principles in Adam Smith's *The Wealth of Nations* (1776), and in the nineteenth century was even seen as the axiom that every political economist should take as a starting point, shows that what at first goes against everyday knowledge can later become part of it.

To a degree something similar applies to Samuelson's model. Since the Second World War, belief in the free market has been regarded more as knowledge than as faith. What Samuelson makes clear is that the market does not always do its beneficial work unconditionally, one example being the transfer of income from one generation to another. By taking us with him into a world in which he addresses this issue of the handing down of wealth and looks at it in a naked form, he is able to call into question the assumption that markets work perfectly. He can then go on to show that a market society needs compulsory forms of cooperation (taxes, state pensions) in order to function efficiently.

7.8 In conclusion

In this chapter, we have discussed an example of a mathematical economic model that can be seen as a test of the intuition of economists that the workings of the market produce the best outcomes for actors in the market, resulting in an optimal allocation of goods and services. Just how deep that intuition lies is clear from the lamentation of a Dutch professor of economics and prominent social democrat in 2001 that the general public really ought now, finally, to take account of the 'truths of economic science', namely that markets lead to efficient outcomes. This chapter has shown that such thinking is not without its problems.

True, Samuelson's model rests on completely unrealistic assumptions, yet it teaches us something about our world. Not for the reason Friedman would give, since he would claim that the reality content of assumptions is irrelevant as long as the predictions are correct. Nor according to the demand that Hausman makes of models, insisting that they must produce hypotheses that can actually be tested. Obviously, the criterion for the usability of Samuelson's model does not lie in its predictive power, since it is clear from the start that his model leads to outcomes that are beyond the boundaries of observation. So what is the value of such models?

For philosopher of science Daniel Hausman, the answer lies in conceptual exploration. Samuelson's model calls into question the meaning of concepts. His analytical exercise reveals the formal implications of the concepts he examines. According to Hausman, Samuelson's model has no empirical consequences, indeed he stresses that it is difficult to see how that model could have empirical implications and he concludes that ultimately it amounts to the same mathematical playing around as the theory of general equilibrium that ran into the sand in the late 1970s.

If we take Samuelson's model to be a thought experiment, we arrive at a different picture. Samuelson constructs a world that is deliberately made different from ours with one exception: transfers of wealth are coordinated through markets. Samuelson demonstrates that in such a world, markets can fail. If we then look at our world, institutions that (for those who believe in the beneficial workings of the market) seemed superfluous suddenly make sense: compulsory pension savings; money as an instrument for transferring value to the future. Anyone looking for predictions, like Friedman, or testable hypotheses, like Hausman, will wonder what is empirical about such conclusions. Samuelson's model neither makes predictions nor represents a real economy. It is a thought experiment that calls into question fundamental intuitions that economists have about how market economies work and, at the same time, it explains why the world in which we live does not, indeed cannot, accord with those intuitions. It shows why our world is not the way we would like to think.

Samuelson formulated his thought experiment mathematically, since he believed mathematics to be the language in which economists needed to express themselves. In this respect, he differs fundamentally from someone such as Robbins, who also sees the economist as a person who understands the world from his or her armchair, but who does so in the form of words rather than mathematics. We have seen from the example of Malthus that a thought experiment does not necessarily have to be formulated in mathematical terms. Conversely, it is not the case that a mathematically formulated thought experiment will have no consequences for our understanding of the world around us. Mathematics turns out to be more than a language; it is an instrument that can furnish insights into the structure of reality.

Notes

1 In the preface to *Foundations*, Samuelson claimed that the subtitle of his thesis was 'The *Operational* Significance of Economic Theory', although the title page clearly reads *Observational*. This makes one wonder just what the verb 'to observe' means in economics.
2 The book was passed by the censor on condition it was sold for the high price of three guineas, but it nevertheless became a bestseller by the standards of the time.
3 This is overlooked by Hausman, who makes Abraham, Isaac and Jacob the protagonists of his explanation of Samuelson's model – exactly the kind of situation Samuelson rules out from the very start.

References

Edwards, Ward (1954). The Theory of Decision Making. *Psychological Bulletin*, 51(4), pp. 380–417.
Hausman, Daniel (1992). *The Inexact and Separate Science of Economics.* (Cambridge: Cambridge University Press).
Malthus, T. Robert (1986 [1798]). An Essay on the Principle of Population, in *The Works of Thomas Robert Malthus*, E.A. Wrigley and David Souden (eds), vol. 1. (London: Pickering).
Samuelson, Paul A. (1958). An Exact Consumption-Loan Model of Interest with or without the Social Contrivance of Money. *Journal of Political Economy*, 66(6), pp. 467–82.

8 Experimentation in economics

8.1 Introduction

In Chapter 7, we learned to see the economist as a theoretician who prefers to steer well clear of empirical research. This does not make his work inconsequential. Although Paul Samuelson made much of the 'operationalism' of Percy Bridgman (who stated that entities can be defined only through the operations by which they are measured), he carefully cultivated the image of the brilliant mathematical theoretician who sent his thought experiments out into the world from the cafés of the MIT campus. Samuelson may have been a very different kind of economist from Friedman, but there was one thing they agreed on: economics was not an experimental science. It was mathematical, but the controlled experiment – since the nineteenth century more or less the litmus test that determined whether or not a given discipline was a science – was not something it could engage in.

It is precisely this consensus that has been called into question fundamentally over the past forty years with the rise of experimental economics. After a cautious start in the 1960s and 1970s, the community of experimental economists has been growing at an increasing rate since the 1980s – to such an extent in fact that even economists who used to think little of this modern trend now perform experiments to support their arguments. Young experimental economists were virtually thumbing their noses at the establishment when they joined forces in the early 1990s to create the Economic Science Association, a choice of name that suggested their work was making economics a 'science' for the first time.

The rise of experimental economics also led to a new pattern of publication. Experimental economists now look to place their research in journals such as *Science* and *Nature*, or in magazines for neuroscientists. The science of economics is clearly in flux, and the self-image of the economist is changing too, but in the 1970s this was not yet the case and it is there that we begin this chapter.

8.2 A dialogue between Vernon Smith and Charles Plott

In the mid-1970s, two pioneers of experimental economics, Vernon Smith and Charles Plott, would regularly set out on fishing trips together in the many areas of natural beauty in the United States. Vernon Smith had been trained as an electrical engineer and became an economist after taking the subject as a minor at university. In the 1970s, he worked at the University of Arizona, becoming one of the first to set up a laboratory for experiments in economics. Charles Plott was affiliated to Caltech, a prominent research university in California. In the early 1960s, Smith had started carrying out experiments in economics and Plott followed in the early 1970s, with experiments in the field of public policymaking among other things. Smith won the Nobel Prize for Economics in 2002, along with psychologist Daniel Kahneman, for his market experiments and more generally for his contribution to the development of the experimental method. In the mid-1970s, experimentation in economics was still far from generally accepted.

We encounter Smith and Plott on a fictional fishing trip in those years. The mood is somewhat querulous. Smith is trying to organize a workshop about experimental economics and he has invited Milton Friedman as a guest speaker. The National Science Foundation (NSF) has promised funding on that basis, but the big fish is refusing to bite. They enjoy a spectacular view of the mountains around Lake Tahoe on the border between California and Nevada. Earlier in the morning, Smith grumbled at length about the fact that Friedman is unwilling to attend the workshop, but now they have been sitting for several hours side by side in silence, watching their floats. They have not made a spectacular catch, but it is good enough to feel pleased about. This evening they will cook the fish near their tent and eat it accompanied by some potato chips, cans of Coca-Cola and a great deal of coffee. They talk loudly above the growl of the outboard motor.

Plott: 'So what did you expect? That Friedman would say yes? You know what he thinks about experiments.'
Smith: 'What did you say? Wait a moment, I'll move over so I can hear you better.'
Plott: 'I said surely you must have known Friedman wouldn't come.'
Smith: 'Maybe I did, but it surprises me all the same. In his essay on methodology he writes that economics doesn't lend itself to experimentation, but he was one of the first to comment on a 1930s study in which Thurstone tries to determine the shape of indifference curves experimentally.[1] So he is at least interested in experiments.'
Plott: 'I'm not really convinced of that, Vernon. Friedman surely didn't believe that you could conclude from Thurstone's study that

consumers actually show their preferences. Those were all hypothetical comparisons between, what was it again, hats and coats?'

Smith: 'That's right. According to Friedman, as an economist you could conclude something meaningful from an experiment only if there was a real incentive for the participants to reveal their preferences. Otherwise what does it mean if someone says he'd be just as happy to have two hats as to have a coat? Experiments in economics would need to involve real payments and rewards, so that participants would show their real preferences.'

Plott: 'Didn't Friedman also believe that market phenomena were far too complex to be captured in experiments?'

Smith: 'That's often what you read about Friedman, but I think the point has been greatly exaggerated. I told you, didn't I, about the experiments Edward Chamberlin did with students during his lectures at Harvard?'

Plott: (*interrupting*) 'Yes, I know that story, you've told it so often and Chamberlin is not Friedman.'[2]

Smith: (*ignoring him*) 'Chamberlin didn't have much faith in the efficacy of market mechanisms and he used experimentation as part of an attempt to show that his own model of monopolistic competition gave a far better picture of what happened in reality than the notion of perfect competition.'

Plott: 'And every time you tell me, you say it was precisely because of those experiments that you got the idea of using experimentation to show that markets do in fact work well.'

Smith: 'Precisely, because of course Friedman was right to say that it's impossible to control for all the factors that influence economic activity. But we don't need to. We only need to create a situation in which we can say with some confidence that individuals act based on a controlled incentive.'

Plott: 'You mean your idea of "induced values"?'

Smith: 'Yes, that too, but there's also Friedman's idea that economic theories give "as if" explanations. Of course in the laboratory you can't fulfil the demands of pure theory, nor can you replicate reality in its entirety – what would you learn from that anyhow? – but the behaviour you observe in a laboratory can lead to outcomes *as if* the theory was valid. As for those induced values, there's an "as if" idea there too, of course. Without game theory I might never have arrived at it, but the idea is really very simple. In situations where people display strategic behaviour, one strategy will win out over another, for example because someone expects to earn more from a particular strategy. In an experiment you want to achieve something like that for individual acts. Supposing I want to use an experiment to look at whether supply and demand in a market lead to equilibrium.'

Plott: 'Surely you don't need game theory for that.'

Smith: 'But game theory and oligopoly theory, which were tested in the lab in the early sixties, did become a source of inspiration, and you know as well as I do that the application of game theory is increasingly routine when you set up an experiment.'

Plott: 'Yes, but...'

Smith: (*impatiently*) '... so to see whether supply and demand reach equilibrium in an experiment, I need to have reasonable grounds for assuming that participants in the experiment are acting according to the supply and demand schedules that I want to test.'

Plott: 'For example by making the reward for behaviour match the behaviour you want to see in the participant.'

Smith: 'And what you then see is that a market achieves equilibrium remarkably quickly.'

Plott: 'The market in your experiment, you mean.'

Smith: 'But no less a market for that.'

Plott: 'I agree there. The standard argument, including Friedman's, against the experimental method in economics is that markets (and economies in general) are so complex that they can only be studied "in the wild". Samuelson says the same thing in that famous textbook of his: economists cannot experiment the way biologists or chemists do, because they can't control the factors that are important. But what's the alternative? Take an econometric model such as those used by Klein or Tinbergen for the United States – they're all about a specific economy that they try to model in every detail...'

Smith: 'Friedman was right, of course, to say that it's an impossible task and that it renders up nothing at all, because the predictions...'

Plott: (*continuing with his own train of thought*) '... and you don't have to, because what you're interested in as an economist are the general principles on which market trading is based, and you can see those far better if you look at simple situations where participants in the experiment are presented with real incentives. It's only reasonable to expect that if a principle operates in complex reality, then it will also operate in a simple situation in a laboratory.'

Smith: 'So an experiment is a model for a real market?'

Plott: 'Right... Or, no, it's exactly as you said just now: in an experiment I create a real market. It's not a model, it's for real. I, as the experimenter, determine the rules of the game. I ensure that the reward structure is such that it would be very ill-advised for a participant not to demonstrate the behaviour I want to see, and I then look at how that market behaves.'

Smith: 'But how do you know that people take decisions in the same way outside the laboratory?'[3]

Plott: 'I don't think that's an interesting question. There are all kinds of markets in which people take decisions. I'm not intending to replicate one of those markets in each and every detail in a laboratory, but I think economists are wrong to see their discipline as a "non-laboratory science" for that reason. I put together a simple situation and then study its characteristics, and I use a simple market as an example, so that I can understand the principles by which markets operate. Whether or not people actually do make decisions based on the so-called axioms of choice theory, say, is irrelevant. Friedman expressed this rather well by saying that if the markets in my laboratory function efficiently, then I can assume that people act *as if* those axioms are valid.'

Smith: 'That appeals to me – but let me play devil's advocate for a moment: you know the Allais paradox, right?'

Plott: 'I'm actually writing a piece about it and similar choice paradoxes with my colleague David Grether. I've heard that two experimental psychologists, Daniel Kahneman and Amos Tversky, are also writing something about the limits of rational choice behaviour, for *Econometrica*, but psychologists, in my view, are very careless in carrying out their experiments.'[4]

Smith: 'Really? I found and still find the work of Fouraker and Siegel in the early 1960s an instance of a good experiment.'

Plott: 'You mean their 1960s experiments on oligopolies and bargaining behaviour?'[5]

Smith: 'Yes, and their 1962 article with Harnett in *Operations Research*. Siegel performed some great experiments about bargaining and collective decision-making. It really is a pity he died so young.'

Plott: 'You're right there. Siegel and Fouraker demonstrate beautifully how an economic theory can be tested with the help of a laboratory experiment. They also show that a combination of economic theory and experimental psychology has a great deal to offer scientists and others who are concerned about finding constructive solutions to social conflicts. But that isn't the sort of experiment I mean when I talk about poorly executed psychology experiments.'

Smith: 'What do you mean then?'

Plott: 'I mean that people such as Kahneman and Tversky test rational choice theory by presenting participants in the experiment with choices that don't really hit them in their pockets.'

Smith: 'It does indeed make a crucial difference whether or not participants act based on monetary incentives. But I want to get back to the Allais paradox for a moment. What Allais shows is that even a great statistician such as Jimmie Savage can fail to make consistent choices in conditions of uncertainty.'

Plott: 'Right. Like most people, he opted for a guaranteed payment of a million rather than gambling on getting five million but changed his mind when chance was involved either way, whereas for a consistent choice he ought to have gone for a million then as well.

The experiments David and I have done looked at just such situations, where participants change their preferences. That is of course a major problem for anyone who believes in rational choice theory as a factual description of human dealings, but David and I certainly don't.'

Smith: 'Doesn't the article that Kahneman and Tversky are writing for *Econometrica* try to repair rational choice theory by means of a weighting function that allows them to take account of subjective decision weights?'

Plott: 'So I'm told, but again, David and I aren't concerned about that, and I don't believe you are either. Experiments of that sort are a disastrous route for economists to go down. Why would we try to make rational choice theory so general that it can explain even inconsistent changes of preference – what use would that be? You have to draw a clear distinction between market experiments and behavioural experiments. The latter are fodder for psychologists, but what we're interested in is market theory. Which brings me back to Friedman...'

Smith: '... it was precisely the problems of economic theory that motivated me to follow Chamberlin and Siegel in developing market experiments. I saw my first market experiment as a simulation of characteristic features of a market, not as a simulation or representation of a specific market. In fact, I tried to exclude any reference to any specific real-world market from the instructions I handed out to my students. The experiment was designed roughly as follows. I divided the participants into two groups, representing supply and demand, and gave them differently coloured cards. Those wanting to buy goods were given a card showing the maximum price they were willing to pay and those supplying goods were given a card in another colour showing the minimum price they were willing to accept. A lower or higher price, respectively, was fine of course. Each participant knew only what the maximum or minimum price was in his or her own case. I did the experiment in an ordinary classroom with my students. That wasn't without its difficulties, but it was handy as well: my students only needed to put their hands up to make an offer, and if buyer and supplier agreed, then the price was noted and they took no further part. This went on until no additional deals were made. We tried to fit as many "market days" as possible into a single session. I changed the schedules of supply and demand and sometimes gave a push to the demand side, sometimes to the supply side, to see what would happen. At the end of the lesson I paid my students. My faculty thought it was amusing and gave me some money for the purpose. The adjustment to equilibrium is remarkable. So remarkable – and that's the strength of this kind of simple market experiment – that you wonder how hard it can be to prove the adjustment process of a market theoretically, if you can see it happen right in front of your eyes in an experiment.'

Plott: 'The market works like a kind of computer – a calculator.'

Smith: 'Yes, and of course it doesn't fulfil the conditions of pure theory at all, but the laboratory market behaves *as if* those conditions have been met.'

Plott: 'Exactly! What you see is that a market is an interplay of individual behaviour and of institutions – it's nonsense to conclude from the fact that individuals violate the rationality postulates of choice theory that therefore markets don't work. Institutions discipline individuals, who therefore act in a certain way as a result. In an experiment, you create and vary institutions to see what the effect will be on the outcome of collective dealings. You use the lessons you draw from that to evaluate other situations.'

Smith: 'In the same way as you can use experiments on rats to learn something about human beings.'

Plott: 'Really? I don't like those kinds of comparisons with other sciences. I believe that the differences are greater than the similarities. Take for example an experiment I'm doing at the moment with my student Jim Hong on the efficiency of the market for the transport of grain and other bulk products on the inland waterways. We're looking at the different effects on market efficiency of privately negotiated prices and what you and I think of as "posted prices".'

Smith: 'Hey, watch out! We need to go left here. Easy does it.'
(*Plott steers to the left and slows down. They are rapidly approaching the beach and the jetty.*)

Plott: 'The rail freight companies want the water freight companies to announce their prices beforehand, claiming the prices will then fall as a result. We've simplified that market by scaling down relevant market factors. So there's one participant in the experiment for every four companies. We've scaled down the time period too, just as you did in your classroom experiment, along with statistical data about the supply and demand structure of the market and several other factors. Afterwards we scaled the results of our experiments back up to the level of the entire market.'

Smith: (*loudly*) 'Slow down! You need to slow down!'

Plott: 'What we were investigating in the laboratory was not a different thing but the same thing; it remains a market, simplified but still a market. And it looks as if we'll be able to conclude from our experiments that the rail freight companies should lose the case. We're curious to find out what the Interstate Commerce Commission will decide about its pricing rules, but we're confident that our experiments shift the burden of proof.'

122 *Experimentation in economics*

Smith: 'So why haven't I seen anything of this experiment yet? Very interesting. I believe it's the first time experimental research has been used for the purpose of policymaking.'

Plott: 'That's right. A first for experimental economics.'

Smith: 'But you surely need to say a bit more about my suggestion that experiments in economics function like experiments on animals. Just as you can carry out experiments with fruit flies to learn something about evolutionary pressure, or experiments with E. coli bacteria or rats to learn something about the development of cancer in humans, so you can use your miniaturized market as an example of processes that also take place on a larger scale and in a more complex environment. Your scaled-down market is not a representation of the real market for water transportation of bulk goods, but it does take on the essential characteristics of it.'

Plott: 'That is indeed the way in which Jim and I are writing about it – I like thinking in terms of the *principles* of a market.'

Smith: 'But even then you don't know for sure whether the results of an experiment are also valid for the real market. You can only be certain by taking the operations you carried out in the laboratory and letting them loose on the outside world.'

Plott: 'So again it's the problem of external validity[6] – I hate that one! (*suddenly animated*) Vernon, we tie up on the left, would you grab that rope near the bow?'

Here their conversation ends, so it is time to take a closer look at the many subjects Plott and Smith have touched upon.

8.3 Traditional views of experimentation in economics

In earlier chapters, we saw that from John Stuart Mill onwards, economists disagreed about many things, but there was striking unanimity in their verdict that economics was not an experimental science. This was also the message of Milton Friedman's famous 1953 essay, which we discussed in Chapter 6. It is certainly true that econometricians compared their way of working with the experimental method, but they generally made that comparison in a metaphorical sense, or with reference to the 'experiments of nature' that made the application of formal probability theory possible; a natural experiment meant drawing lots from an urn full of possibilities. In Chapter 7, we drew a parallel between Samuelson's building of models and 'thought experiments', but that was a term we applied to it ourselves, not one Samuelson himself used. As for his famous 1947 textbook *Economics*, which sold some four million copies, it was not until the editions published in the 1970s that the experimental method was even mentioned, and then only in a negative sense, in a way that corresponded with Friedman's attitude: 'Economists ...

cannot perform the controlled experiments of chemists or biologists because they cannot easily control other important factors' (Samuelson and Nordhaus, quoted in Guala, 2005, p. 3). Canadian economist Richard Lipsey, in his own textbook (*An Introduction to Positive Economics*, 1995), which was particularly successful in Britain, wrote in a similar vein that economics was a 'non-laboratory science', because it would rarely if ever be possible to carry out controlled experiments on an economy.

All these economists essentially adhered to the reasoning found in John Stuart Mill's 1836 essay: economics deals with an extremely complex reality, in which 'disturbing causes' mean that no law-like regularities are visible; the science of economics is a science of tendencies. Whatever its method might be, it is not the experimental method. The entry for economics in the *Encyclopaedia Britannica* actually stated that economics was not an experimental science, because there was no laboratory in which economists would be able to test their hypotheses.[7] Quite some ground separated economics from the controlled experiment, the litmus test of good science in other fields.

This simple observation was in accordance with the everyday intuition of the economist: markets involve so many different players that fitting a market into a laboratory is inconceivable. Economics was seen as lacking a place – the laboratory – where experiments on its object of study could be performed. If there was a 'place' for experiments then it was the mathematical model, and from the 1970s onwards there was an increasing tendency to refer to changes in the parameters or the specifications of the model, or to other similar changes, as a 'model simulation' or 'experiment on a model'. Aligning themselves with Robbins' definition of economics, many economists satisfied themselves with a notion of economics as a discipline looking at just one aspect of human dealings. People have so many different motives for acting that economics, they felt, investigates individuals only insofar as they act to optimize outcomes for themselves, and in that sense rationally.

The fictional conversation between Vernon Smith and Charles Plott makes clear that the introduction of the experiment into the discipline of economics represented a radical change to its working method. We should not be surprised, therefore, that the 2002 Nobel Prize for Economics was awarded to Vernon Smith, the engineer who was interested in the rationality of markets, and to Daniel Kahneman, the psychologist who was interested in the rationality of individuals.

8.4 Market experiments and behavioural experiments

Nowadays, a distinction is made between behavioural economists and experimental economists. This is confusing, since both groups perform experiments. It would make more sense to distinguish between market experiments and behavioural experiments. Vernon Smith's supply and demand experiment and indeed Charles Plott's research into the transport market for bulk goods on the inland waterways are examples of market experiments, but even a distinction between market experiments and behavioural experiments tends to conflate a method with a field of study. From John Stuart Mill onwards, economists were sceptical about the use of the experimental method in economics, yet such scepticism should not be confused with the perceived difficulty of applying the experimental method to a specific topic, in this case to complex economic phenomena such as markets. No one denied that experiments could be carried out on individuals, to study their behaviour in experimental conditions.

In a market experiment, participants are asked to perform a specific, well-defined task according to a number of clearly articulated rules that describe how the market works and which 'moves' are allowed or forbidden. The reward structure is designed such that the participants in the experiment have a strong incentive to behave according to the rules of the game. (This, in a nutshell, is what Vernon Smith's theory of induced preferences aims to accomplish.) After a number of test rounds that serve to ensure everyone fully understands the rules, the experiment itself begins. It usually takes one or two hours, but sometimes, especially in the early years of such experiments, the whole exercise might take several days. Participants are usually, although not necessarily, members of the faculty's student population. The experiment involves tokens that can be exchanged for real money at the end. How much an individual earns depends on his or her decisions but, in most cases, there is a minimum amount that all participants receive.

Game theory, the mathematical theory of strategic decision-making, is currently used almost as a matter of course in designing experiments of this kind. Often there are several possible outcomes, and there is no way to determine analytically what the results will be. Experiments serve to show which outcome is achieved. By changing the rules of the game, or by altering other variables (such as economist/non-economist, man/woman and so forth), it is possible to see the effect of specific variables on outcomes. One type of market in particular, the auction, has become popular as a result of the rise of experimental economics. Because of experiments by economists, it is now possible to develop markets (or auctions) for products for which there was previously no market, or to give markets a structure that has never been found 'in the wild', in other words to create 'synthetic markets'. One famous example is the FCC (Federal Communications Commission) auction in the United States, another the UMTS (Universal Mobile Telecommunications System) auctions of the 1990s in Europe, in which radio frequencies were sold to telecommunications companies, but there are plenty of other examples, from the auctioning of landing rights at airports and the allocation of places in the permanent Space Station to markets that match donated kidneys with transplant patients. Nowadays, auctions are fairly regularly designed in the laboratory and the design is then sold to governments or businesses.

In behavioural experiments, the central concern is not the efficiency (or any other characteristic) of a market but rather the decision-making process engaged in by individual participants. Do individuals behave rationally? What part is played by norms in choice behaviour? What is the role of emotions in decision-making? As a result of the 1970s work of Amos Tversky and Daniel Kahneman in particular, behavioural experiments made the transition from the field of psychology to economics. For a long time, various sorts of choice paradox lay at the heart of this kind of research, concerning robust experimental results in which participants violated one or more postulates of a specific version of rational choice theory: expected utility theory. By robust I mean that the results were reproduced, time and again, in different laboratories. These are results that do not go away. In many such experiments, uncertainty has a crucial place; people turn out to be astonishingly bad at applying formal probability theory. This was demonstrated even by one of the founders of modern probability theory, Jimmie Savage, who like many others fell into a trap set for him by French economist and mathematician Maurice Allais shortly after the war.

Without going into too much detail, Allais' trap can be summarized as follows. Imagine that someone chooses a safe bet (A) over an uncertain gamble (B). Allais ingeniously altered the likelihoods of A and B, but he did so in such a way that the relationship between the two remained precisely the same. Anyone who makes a consistent choice ought therefore to opt for A over B in the new situation as well, but typically they do not. Even Jimmie Savage changed his choice and opted for B over A, despite the fact that according to the very axioms of choice that he had formulated himself, he ought not to have altered his decision. Since then, other choice paradoxes have been described, many of which can be 'explained' in the economics laboratory, but the Allais paradox remains particularly tenacious and within the science of economics it has attained the status of what historian and philosopher of science Thomas Kuhn calls an anomaly – a problem that cannot be resolved within the paradigm of present-day economics.

Behavioural economists take on these problems of choice. They investigate (among other things, not exclusively) the many factors that can help to explain the inconsistencies in decision-making behaviour that emerge from laboratory experiments, or more generally how decision-making behaviour departs from predictions made for it by economic theory. As I have already suggested, they might look at the role of emotions in decision-making, or the impact of moral convictions or cultural factors. A standard example is the outcome of what is known as the dictator game, in which one participant is asked to share out, as he or she thinks fit, a sum of money, perhaps 100 euro, between him- or herself and another participant. Economic theory says that the person who is initially given the money will keep the entire sum, but this is typically not the outcome. On average, the 'dictator' will give the other person between 30 and 40 euro. Another example is the contribution to so-called public goods, where again the outcome differs from the theoretical prediction. Those doing this kind of research have an increasing tendency to try to find a connection with the neurosciences or the medical sciences more generally; the question then is whether divergences from predictions made by economic theory can be attributed to the way our brains are structured or whether they might arise from the presence or absence of specific hormones, such as oxytocin or dopamine.

One important concept in behavioural experiments is framing: how is a problem or question formulated? Its significance was clear from the Allais paradox, and here too experimental research has produced consistent results. Whether a situation in an experiment is presented as one of competition or cooperation affects the outcome. Whether someone is automatically expected to be a donor or is left to choose has a radical effect on the proportion of donors to non-donors. There are many examples of this kind, and over the past few years they have found their way to policymakers and to a broad audience.

Market experiments and behavioural experiments are not necessarily mutually exclusive, but they do presume different experimental cultures and usually, although not necessarily, different political agendas. So even though it is not right to draw a firm distinction between them, in practice there is a divide between the two experimental communities. The argument by Vernon Smith that a market organizes human rationality even if individuals violate the axioms of choice theory implies a political preference for the kind of society that depends on well-organized markets, with people and businesses disciplined not by governments but by the market itself. Behavioural economists will instead tend to allocate an important role to governments in organizing choices, or in restricting the room for choice, because they conclude from their experiments that limiting choices leads to better outcomes. Both in market experiments and in behavioural experiments, the design of the environment is important (especially where it concerns 'institutions' in market experiments and 'framing' in behavioural experiments). This, in itself, is not a reason for there to be any difference in principle between the two sorts of economist, but in the American context the debate over the healthcare reforms put in place by President Obama shows that a change in emphasis leads in practice to an unbridgeable difference of opinion. Current controversies about the efficient working of financial markets are approached (and answered) differently by the two camps. Ideas about what constitutes a 'good' experiment are therefore not independent of the political preferences of experimental economists.

8.5 From classroom experiments to the laboratory

Vernon Smith stressed that his interest in the experimental method was fostered by his teacher at Harvard, Edward Chamberlin, but it was also encouraged by the work of psychologist Sidney Siegel. Initially, Smith tried to use experimentation to solve theoretical problems. His supply and demand experiment in 1960 concerned just such a problem. With the formalization of economic theory, it turned out to be surprisingly difficult to model the dynamics of a market satisfactorily. How does a market reach an equilibrium price – if it does so at all? Any first lecture on microeconomics is likely to involve the drawing of supply and demand curves, which indicate the equilibrium price, but it remains unclear what the 'market mechanism' is by which this equilibrium price is reached.

Anyone who has ever reproduced Vernon Smith's experiment will have been amazed by the speed with which participants are able to 'find' the equilibrium. Smith performed his experiment in a classroom, and all participants were allowed to call out the price they were offering or were willing to accept, but even in the closed-off spaces that are a feature of economics laboratories today (see Figure 8.1), where participants type their responses into a computer terminal and the software presents their answers to the experimental economist on his screen, outcomes quickly converge to equilibrium. It seems the invisible hand does actually work – as long as you design it properly. Without any deliberate desire on the part of the participants, market equilibrium is reached within a couple of steps.

Figure 8.1 Social Science Experimental Laboratory (SSEL) at Caltech, Passadena. The partitions are to ensure that test subjects cannot see anyone's response but their own.

Source: Photograph by Andrej Svorencik.

128 *Experimentation in economics*

Let us look at Vernon Smith's much-discussed 1962 article in the *Journal of Political Economy* to see that (not how) the market converges to equilibrium (see Figure 8.2). The left half of the image shows the reservation prices, as given on cards that the participants receive at the start of the experiment (or at the start of each new round of the experiment). A reservation price is the price above or below which each participant has been asked not to buy or sell. Those taking part know only their own minimum or maximum price and are instructed not to communicate it to other participants. The graph of these reservation prices shows the stepped structure typical of discrete changes in price (the number of participants is limited, so the number of price steps is limited) and it has become an icon of the market experiment. The grey area shows all the possible prices (for example, no one will accept less than a price of 80 dollar cents, so a lower price is not possible). The right half of the image shows the actual adjustment process during each round of the experiment. It is remarkable that, almost from the start, the prices agreed are close to the theoretical equilibrium price.

In Edward Chamberlin's classroom experiments, his students privately searched for other student participants to trade with. Vernon Smith carried out similar experiments and, in his case, prices were called out and written on the blackboard so everyone could see what was happening

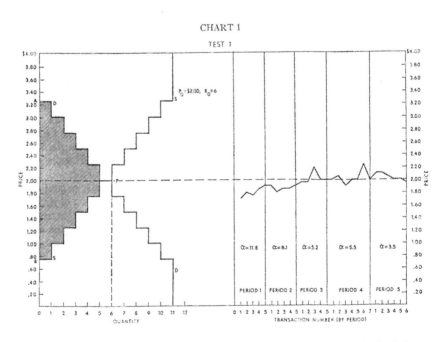

Figure 8.2 Diagram of experimentally generated supply and demand schedules.

Source: Vernon L. Smith (1962). An Experimental Study of Competitive Market Behavior. *Journal of Political Economy*, 70(2), p. 113.

in the market as a whole. Chamberlin and Smith were therefore using different trading rules. Experimental economists came to refer to such trading rules as 'institutions' and much experimental work has gone into the investigation of different institutions in this sense on market outcomes. Smith happened to be extremely good at perceiving the consequences of specific rules for the outcome of an experiment. Economic theorists initially showed little interest in the kind of experimental work done by Smith, arguing on theoretical grounds that changing the rules should not make any difference to market outcomes. If asked, they recall very clearly the moment when they were suddenly converted to his way of thinking, such as when Smith performed an experiment unsuccessfully, just as they had expected, but then, with a small change to the rules, managed to create a situation that converged to market equilibrium in the blink of an eye.[8]

Let us now look in more detail at the joint study by Charles Plott and his graduate student James Hong, in order to gain a firmer grasp of what is at stake in such pricing-rule experiments. We first need to discuss a joint experiment by Vernon Smith and Charles Plott on what has become known as the 'posted-price effect'.

8.6 Experimentally establishing the 'posted-price effect'

In 1978, Charles Plott and Vernon Smith reported on experiments that compared two different pricing institutions. Their paper followed an earlier article by Fred E. Williams, published in 1973. Williams, a graduate of Purdue at a time when Vernon Smith had already left for another university, expanded on one of Vernon Smith's early market experiments in the 1960s, in which Smith examined the 'haggling and bargaining process' in the market when sellers or buyers posted their prices in advance.

Williams considered a number of different offer and bidding schemes as the means of investigating what he thought of as the 'treatment variable' of his experiment – whether sellers or buyers posted prices in advance – and concluded that the form of bidding was an important 'institutional condition'. This was a surprising finding according to the standard supply and demand framework, since institutional settings were not thought to matter. Taking a close look at Williams' paper, especially at how he introduced trading in multiple units per market period, Plott and Smith concluded that he had introduced a time element into the design of the experiment, which significantly altered the pricing rule that Smith had originally considered. Smith had made a loose connection between the haggling that took place in his experimental market and retail markets in which prices are posted competitively, a practice that Williams interpreted as price leadership by suppliers in the market.

Williams took Smith's lead by introducing his paper using similar references to familiar markets, such as retail markets, that offer their goods at stated prices. But whereas in Smith's experiment sellers (or buyers) could adjust prices every time a unit was offered during a trading period (which made 'posting' the equivalent of 'announcing'), in Williams' experiment prices were set at the beginning of each experimental market period and could not be changed thereafter. Plott and Smith conjectured that Williams' results had no bearing upon the price leadership of buyers or sellers, but that they did have a bearing upon what came to be known as the 'posted-price effect'; that is, the fact that prices were higher (or lower) than the theoretical equilibrium price if trade was executed on prices that were fixed by sellers (or buyers) for the entire duration of trading periods.

They were led to make this conjecture by perceived similarities between the posted-prices rule and two different kinds of auction that, when theoretically and experimentally examined, showed similar differences in results (the discriminative auction where buyers bought at their own bid price and the competitive auction where buyers bought at the lowest bid price). Plott and Smith gave real-world examples: the auction of US Treasury Bills for the discriminative auction and the French auction of new stock issues for the competitive auction. They expected similar results for the discriminative auction as for posted prices, because buyers' behaviour would be affected by considerations that were, in the end, the same under both regimes. Economic agents would behave differently in a posted-prices regime from the way they behaved in a competitive auction.

Based on this interpretation, Plott and Smith designed a new experiment in which they compared the process of achieving equilibrium and efficiency under two pricing rules: posted prices (which could not be changed during a market period) and an oral auction (where prices could be adjusted continuously). They spent a good deal of time reinterpreting Williams' experiment, taking care to design their own experiment in such a way that they could 'replicate' or 'reproduce' his results (they used the two expressions interchangeably). They drew attention to one of his diagrams that showed the seller's posted offers as being consistently higher than the buyer's posted bids. Diagrams based on their own experiments were consistent with this finding.

If they could reproduce Williams' result in such a way that it was in line with their conjecture, then they would have shown that posted prices will always lead to a higher or lower equilibrium than the oral auction, depending on whether suppliers or buyers post prices in advance. They stressed that their experiments did not test any formal theory, rather they were aimed at producing a 'rigorous empirical foundation' for the development of such a theory (Plott and Smith, 1978, p.133). The word 'rigorous' pertained to the way data were produced in an experimental setting; the data were 'rigorously empirical' because they were produced under the control of the person carrying out the experiment. Plott and Smith did not mean by this that one could now speak 'rigorously' about differences in price formation at supermarkets or in US Treasury bond auctions. They found that the posted-prices institution was less efficient than the oral auction, and that, of the two, it resulted in lower equilibrium prices.

On several other occasions, Plott and Smith emphasized that experimentally constructed markets are no less real than markets outside the laboratory because 'real people earn real money for making real decisions about abstract claims that are just as "real" as a share of General Motors'; there is real trading by real agents taking place in these markets, and there is real price formation.[9] This is as true for Plott and Smith's experimental comparison of pricing institutions as it was for Williams' experiments. But a market can be real and yet abstract. Such a market embodies the generic characteristics of actual, real-life markets. The posted-prices rule in the Smith–Plott experiment refers to a pricing rule used in the retail market, but their rule lacks the characteristics of specific retail markets that may make it play out differently in reality. Also, real people participating in a generic experiment may differ from decision-making entities in the outside world, whether they be individuals or collectives.

One way to express the difference between experimental markets and markets in the real world is in terms of simple versus complex. This is in line with how economists, as we have seen, traditionally thought about the difficulty of experimental research in their field: real markets are too complex. Experiments in economics can themselves become fairly complex, however, and it may be better to replace this distinction with one between generic and specific. In generic experiments, there is no reference to how institutional rules play out in the real world, whereas specific markets do try to take concrete circumstances into account. The experiment by Hong and Plott on the American freight transport market is an example of an experiment in which the experimenters tried to include the actual circumstances of a specific market in a laboratory setting. To put it another way, this experiment was both real and concrete, not real and abstract; it was not generic but specific.

8.7 Hong and Plott's paper on filing policies for inland water transportation

Hong and Plott performed an experiment explicitly targeted at a specific, real-world market: the market for the transport of bulk products such as grain on US inland waterways. In the 1970s, and indeed earlier, prices in this market were privately negotiated between the parties involved. Rail freight companies felt that this pricing policy was preventing them from entering the market. They argued that market efficiency would increase if all water freight companies had to post their prices publicly, so that buyers could make an informed choice between the competition. An economist would intuitively feel this was right: posted prices increase market transparency and therefore market efficiency. The companies also argued that existing pricing policies favoured large firms over smaller ones, and that the reverse would be the case for the posted-prices rule.

Plott's attention was drawn to this dispute by his former fellow student John W. Snow when the experiments on the posted-price effect were already underway.[10] The results of the joint work by Plott and Smith made the credibility of the argument by the rail freight companies much less straightforward. With the support of the US Department of Transportation and the National Science Foundation, Plott and his student James Hong published their results as a working paper in the spring of 1978.[11] The paper was rejected for publication by the *Journal of Political Economy* in the autumn of that same year, and it was a while before Plott could publish it elsewhere. The rejection by the *Journal of Political Economy* reflects the fact that the Hong–Plott study was seen as too specific for a journal with a theoretical focus, but Plott's submission of the work to this journal in the first place indicates that he did not consider the paper to be merely a piece of applied research. He clearly felt it was interesting from a theoretical point of view as well.

As we have seen, Plott and Smith investigated two different pricing rules in a generic setting. They did not look at implementation of those rules in markets 'in the wild'. They included only casual observations about specific markets, as pointers and to indicate why their study was of interest in a broad spectrum of cases. In the case of the freight transport market, however, a concrete situation lay behind the design of the experiment, its execution and the discussion and interpretation of its results, as is indicated by the titles: 'Rate filing policies for inland water transportation: an experimental approach' is considerably more specific than 'An experimental examination of two exchange institutions'. The first paper deals with experimental research that has a bearing on a specific market located in time and space, the second with experiments on abstract pricing rules.

This difference in orientation, specific versus generic, had consequences for the material configuration and interpretation of the experiment. Hong and Plott could not be content with casual references to real-world situations that made use of the posted-prices rule or some other pricing rule; they had to provide a detailed discussion of how the real-world set-up of the inland freight market could be converted to laboratory dimensions, and the reverse. This involved looking at how the target market could be scaled down to laboratory proportions and how the results of the experiment could then be scaled up again to match the world outside the laboratory. Nowadays, we would consider these issues as concerning the external validity of experiments – the question of whether experimental results also hold true outside the laboratory setting – but Hong and Plott explicitly presented their discussion in terms of scale, and in what follows we will consider the usefulness of that vocabulary.

In the mid-1970s, the term 'external validity' was not used by experimental economists such as Plott and Smith. Its history can be traced back to experimental psychology rather than economics, where the term was introduced only in the second half of the 1990s. Discussions about external validity in experiments by economists are linked to the behaviour of participants under laboratory conditions as compared to the behaviour of individuals 'in the wild'. But one of the issues in Hong and Plott's laboratory experiments is the ambiguity of the notion of the economic agent 'in the wild' and it is precisely here that questions of scale are pertinent.

8.8 Scaling down the freight transport market

How did Hong and Plott manage to create a credible laboratory representation of the market for bulk freight transport by water? 'Within recent years', Hong and Plott wrote, 'several programs of rate publication have been proposed' in this context (Hong and Plott, 1982). One proposal, favoured by the rail freight companies, was that rates should be filed with the Interstate Commerce Commission 'at least fifteen days before a rate change is to become effective'. Because this was not the pricing policy currently in use (which depended on individual negotiations), Hong and Plott attempted to incorporate 'several prominent economic features' of the bulk freight industry into an experiment designed to test the efficiency of different pricing rules. 'Incorporating prominent features' meant they had to scale this particular market down to laboratory dimensions. The results were subsequently scaled up again to market level. Table 8.1 gives a schematic overview of the most important issues involved in the design and outcome of the experiment.

First, Hong and Plott decided to study only part of the inland water freight market, the market for grain transport, and indeed only part of that market: on the upper Mississippi and the Illinois Waterway, during the autumn of 1970. The reasons for this decision had to do with pragmatic concerns about data availability, but also about the characteristics of the grain market itself. Grain was a homogenous commodity and its units could easily be scaled down to experimental proportions. Grain accounted for a large part (70 per cent) of the market they were examining. Grain also made up a significant proportion (35 per cent) of the entire market for bulk transport by water in the United States (excluding the Pacific Coast). They chose 1970 for reasons of data availability and because it showed 'typical historical trends' – a point upon which they did not elaborate further.

Table 8.1 A comparative overview of the Hong–Plott experiment

Trading procedures	Posted prices		Private negotiations	
	Experiment	Target market (proposal)	Experiment	Target market
Start of trading period	Sound of horn	Submission of prices to Interstate Commerce Commission	Sound of horn	Continuous
Who is trading	Experimental subject	Firm	Experimental subject	Firm
Suppliers	22	±93	22	±93
Buyers	11	Not specified	11	Not specified
Supply curve	Composed of large, small and medium traders, rotating schedules	Composed of large, small and medium traders, flexible entry in the market	Composed of large, small and medium traders	Composed of large, small and medium traders, flexible entry in the market
Demand curve	Composed of large and small traders handed to subjects. Schedules rotate between subgroups of subjects (large, small), shift of curve after ## rounds and after ## rounds shift back	Composed of large and small traders, upward shift in demand (curve) in October and November	Composed of large and small traders handed to subjects. Schedules rotate between subgroups of subjects (large, small), shift of curve after ## rounds and after ## rounds shift back	Composed of large and small traders, upward shift in demand (curve) in October and November
Shape of curves	Published sources, personal judgement of experimenters	Published sources	Published sources, personal judgement of experimenters	Published sources
Payment	Yes	Yes, but not specified how	Yes	Yes, but not specified how

Trading procedures	Posted prices	Private negotiations	
Trading period	12 (15) minutes	12 (15) minutes	Continuous
Where	Offices in Baxter Hall Caltech	Office	Office/don't know
When	7–10pm	7–10pm	Office hours/don't know
Who	33 subjects – housewives, engineers, faculty members, secretaries, graduate students in engineering, business and law, some from Caltech	33 subjects, housewives, engineers, faculty members, secretaries, graduate students in engineering, business and law, some from Caltech	Professional traders in the service of firms
	14 days		
	Office/don't know		
	Office hours		
	Professional traders in the service of firms		
Contact between trading agents	Not permitted, at refreshments machine	No	Probably yes
	Not permitted?		
Collusion	Forbidden	Forbidden	Forbidden/don't know
Trading technology	Telephone	Telephone	Telephone/oral communication
	Forbidden		
	Not specified		
Rate filing institution	Experimenter	–	–
	Interstate Commerce Commission		

Note: the experimental intervention involves changing the pricing rule from private negotiation to posted prices. This change was proposed as a policy change by the Interstate Commerce Commission, after complaints by US rail freight companies about unequal terms of competition that they claimed favoured the water freight companies.

So to bring the inland water freight market into focus, Hong and Plott chose one specific part of that entire market, which was then scaled down to experimental proportions. In an undated working paper, they discussed the comparison between the laboratory and the target market, in terms of both simple versus complex and small versus large:

> Clearly, we are unable to recreate the industry with all its complexities. The best we can do is to create a small market the essential features of which are similar to the essential features of parts of the industry. We then hope to learn something about the more complex industry by studying the behaviour of the simple one.

But the simple–complex and small–large dichotomies do not concern the same kinds of problems. In the former case, issues related to control within the experiment are important, in the second the central concern is not one of control but rather of the legitimacy of drawing inferences from a small-scale world and relating them to the full-scale world. These latter issues are clearly a matter of external validity, specifically where they concern downscaling and upscaling.

The most important parameters that needed scaling down to laboratory dimensions were those that related to supply and demand functions. The limited availability of statistical data precluded the use of standard econometrics so, in reaching a reasoned assessment, Hong and Plott took recourse to existing studies on the shape of supply and demand curves for the transport market for grain, studies on the transport of other bulk products and their own expert opinion. They concluded that the demand and supply curves were both relatively flat. They also looked at the number of suppliers of transport in the market and whether these were large, medium or small. Their estimates were derived in part from reports and empirical papers on the barge industry.[12] Demand for transport in the target market increased in October and November, because of summer harvests, and decreased again afterwards, and this provided additional information on supply elasticity.

Based on this range of evidence and on reasoned guesses, Hong and Plott designed experimental supply and demand schedules that were distributed to individual participants during trading rounds in their experiment. The supply schedules not only scaled down market supply to experimental units, they also took into account the distribution of large, medium and small suppliers in the market. For example, two participants received schedules that together represented a market share of 29 per cent, whereas twelve received schedules that amounted to only 19 per cent. To capture the seasonal rise in demand in late summer, demand schedules increased after a few trading rounds and fell back again afterwards to the initial levels.

The design of the experiment was as follows. Thirty-three participants, mostly Caltech postgraduate students, Caltech staff, housewives or engineers, were split into two groups, suppliers (twenty-two) and buyers (eleven), and assigned different trading schedules. Hong and Plott included a diagram that showed demand and supply in the experimental market (see Figure 8.3). The numbers below the curves refer to the individual supply and demand schedules that were handed out to participants. Since the aim was to evaluate the difference between the posted-prices rule and private negotiations, rounds of the experiment were held after office hours so that there were sufficient empty offices available to place the participants in separate rooms.

Private negotiations were carried out by telephone, in trading rounds lasting twelve minutes (fifteen for the first three practice rounds). All participants had trading schedules (their own reservation prices) and a list of names and telephone numbers on their desks, and they were free to call whomever they wanted. To replicate posted prices, suppliers privately announced their trading prices to the experimenter, who distributed a photocopy of all prices to all buyers as the basis on which they would make their decisions. Outside of the rounds of the experiment, discussion of prices was forbidden (even though participants were able to chat during brief breaks between rounds).

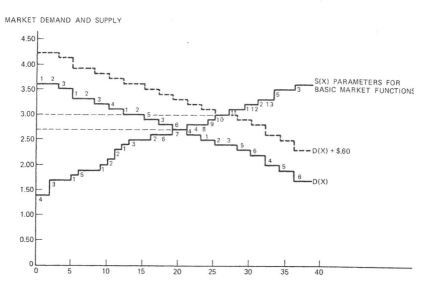

Figure 8.3 Experimental market demand and supply schedules.

Source: James T. Hong and Charles R. Plott (1982). Rate Filing Policies for Inland Water Transportation: An Experimental Approach. *The Bell Journal of Economics*, 13(1), p. 5.

138 *Experimentation in economics*

The instructions read to the participants were the same as those in Plott and Smith's joint study, although of course altered with regard to market organization despite the fact that the setting in empty offices with telephones was markedly different. The instructions also adhered to Smith's recommendations in the sense that the market was described only in generic terms; only the experimenters, not the participants, knew about the specific market concerned. An important aspect of economic experiments is that participants can make real money from their performance. Each trading period started and ended with the sounding of a horn.

The experiments took place over four evenings, between 7pm and 10pm, in four sessions, two for the posted-prices rule and two for the private negotiations. Each session had a different supply curve. There were twelve trading rounds per session (except on the first evening, for reasons explained in the paper but not relevant here). On all four evenings, the demand curve moved outwards for a few trading periods and then moved back. The experimental results were evaluated for market efficiency and for distributional characteristics.

8.9 Scaling up

It emerged that the market with the posted-prices rule performed less efficiently than the market that involved private negotiations between participants. The posted-prices rule also worked to the disadvantage of small suppliers. These experimental results were subsequently scaled up to real-world proportions using the following metric:

> One participant = 4 firms
> One period = 2 weeks
> One unit/period = ½ tow
> One tow = 19.5 million ton-miles of grain per month
> One tow per month in study region
>   = 1 boat's capacity in study region
>   = 5 boats' capacity in all regions

Hong and Plott presented the scaled-up results in a diagram comparing market data for the target market with their experimental results (see Figure 8.4). At first glance, it seems that only the scale on the horizontal and vertical axes was changed, but in fact that was not all that happened. Experimental results were paired up and averaged out over experiments with the same market organization and averaged over two periods in order to convert from fortnights to months. Monthly data for the target market, together with the averaged data for the two experimental market organizations, were plotted on the diagram.

The supply and demand schedules for the experimental market were known, but their shape, and therefore also their shape in the scaled-up diagram, was based on the various sources of information Hong and Plott had used to create the experimental schedules in the first place. The diagram for supply and demand curves in the target market was based on conjecture. It shows that the negotiated and target market prices and volumes were close to the scaled-up experimental equilibrium for most months, with the exception of October and November, when the demand curve shifted outwards. Hong and Plott argued that trading volume under the posted-prices rule was on average less than for the privately negotiated prices, while prices were on average higher and market efficiency lower. They added that the apparent closeness between the values of the posted prices and the hypothetical equilibrium for the shifted demand curve was illusory, resulting from the 'cumbersome and slow' adjustment process under the posted-prices rule. In fact, the posted prices were 'excessively high' when experimental demand increased. 'Cumbersome and slow' referred to the experimental implementation of the posted-prices rule, which involved the collection of supply prices between trading rounds and their copying and distribution to all buyers, a process that had to be repeated for every fresh round.

8.10 The impossibility of altering the scale of people and objects

We have been discussing Hong and Plott's market experiment as if it were indeed an experiment. But was it? Or was it a simulation of market processes? In the early 1960s, the vocabulary used by economists in this connection was still entirely amorphous. Vernon Smith refers to his early market experiment both as an experiment and as a 'simulation'. Others, too, used concepts such as simulation, model and experiment interchangeably. Then there is the concept of the 'business game'. The two notions, simulation and model, were used to indicate the representation of an economy without saying anything about the medium of representation. Simulations and models could be mathematical representations but, in the early 1960s at least, there were simulations that used real people to represent, for example, market or business processes. Whereas in those years the concepts of experiment and simulation were used interchangeably, as time went on the word 'experiment' came to be reserved for those situations in which researchers carried out a controlled intervention, whether in a material or a virtual system. As we shall see in more detail in Chapter 9, it is possible to perform an experiment on a mathematical model, for example.

In the instructions read out to the participants, the experiments by Plott and Smith and by Hong and Plott were presented as 'market simulations'. But whereas Plott and Smith's participants were told that the experiment was 'going to simulate a market' in which some participants were buyers and others sellers, the same information was conveyed in a substantially different situation in Hong and Plott's experiment. Their experimental market was not a generic market of buyers and sellers but a specific market, of which the experimenters had tried to capture the most important characteristics. No hint of this was revealed to those taking part in the experiment. As for external validity, it is clear there is no one-to-one correspondence between the design and material configuration of Hong and Plott's experiment and their target market. The difficulty of establishing this correspondence is where scale issues come in. These emerge when there is a need to scale down the time things take in the target market to the dimensions of the laboratory. They also concern the material configuration of the experiment and the role of its participants.

Let me first address the role of participants. When Plott and Smith compared two pricing rules and concluded that the posted-prices rule resulted in lower bid prices than the bid auction, they ensured that participants in the experiment expressed behaviour that, although constrained by the design of the experiment, left them free to act differently with different pricing rules. The behaviour of the participants provided information about the efficiency of the two pricing rules. Choice behaviour was also constrained by the length of the experimental trading period but, within the design of the experiment, no special significance was attached to the timeframe. Individual participants were therefore taking decisions under similar conditions in both experiments, and this individual behaviour was informative on the topic that Plott and Smith's experiment was intended to examine: how well two different pricing rules compare in terms of efficiency. Were this not the case, we might just as well run a computer simulation on a model and compare outcomes.

The experiment by Hong and Plott is different. A human being in their experiment does not unambiguously correspond to a human being in the target market. Participants play at being traders within firms (despite being unaware of this), so there might seem to be a one-to-one correspondence between their behaviour in the experimental market and that of their counterparts in the target market. But they play a more complex role as well. Hong and Plott scaled down a number of firms to correspond with human traders in the market: one participant was equated to four firms. Does this mean that Hong and Plott believed they could map, one-to-one, the choices made by one participant onto those made by four firms – as if those four firms were acting in tandem? Do four firms collude around a telephone to make market decisions of the kind a single human being makes in an experiment? Of course not. But the issue of how a participant in an experiment 'expands' to the scale of the target market does not arise in the case of the experiments by Williams, or by Plott and Smith. That it does for the Hong–Plott study means that we cannot take the behaviour of the participants in their experiments at face value. We need to consider whether the design of the experiment makes individual behaviour informative about the behaviour of firms in the target market. Hong and Plott use the word 'simulation' to indicate the relationship between their experiment and the target market. To put it another way, the design of the experiment needs to equip participants to play their role in the experiment as stand-ins for firms in the target market.

The telephone, empty offices and paperwork on the desks served as such equipment. Hong and Plott tried to create a situation for their participants that was identical to trading situations experienced by employees of firms in the target market. The trading situation ensured that participants could privately haggle over prices by phone in an empty office, just as negotiators working for companies contacted clients by phone. But time makes a difference. Negotiating over a longer time may elicit different behaviour than will arise when a person taking part in an experiment has to come to a decision within twelve minutes, without the option of, for example, consulting his or her colleagues. While the telephone setting appeared to replicate the trading situation in the target market, scaling down negotiation time to twelve minutes in fact changed the role of the telephone from a communication device to a device that constrained participants' potential to communicate and hence restricted access to information, thereby loading pressure onto participants that was absent (or at least absent in this particular form) in the target market. Hong and Plott reported that telephone lines were 'frequently busy' during the experiment and that seeking information by phone meant foregoing trading opportunities. Brian Binger, one of the participants in the experiment, remembers:

> Each trading session was signalled by a blast of an air horn, and then we were to dial other participants and try to make trades, and one of the things that I recall specifically is that we soon learned that it was very difficult to get through on the telephone system. So, I think most of us adopted the strategy of dialling the first four digits of a trading partner and when the air horn was blown, we dialled the last digit and were able to get through and make a trade with that person without doing any particular searching. And I always wondered whether that technology impeded the ability to search for better alternatives.
> (Quoted from Svorencik, forthcoming)

Hong and Plott's way of putting pressure on participants in the experimental context was part of their deliberate effort to capture features of the actual industry, such as the cost of searching for information and contract opportunities. But dialling the last digit on hearing the blast of a horn was a feature of the experiment, not of the target market. The communications technology used inside and outside the experiment, the telephone, was identical, but placing the same technology in a reduced timeframe made the telephone a different instrument in the experiment than it was in the target market, where it was an enabling device for those needing to contact trading partners. In the experiment, it was a more complex device that constrained the contacting of trading partners and embodied search and opportunity costs.

While the scaling down of time changed the role of the telephone, time also had an important effect on the posted-prices procedure. Hong and Plott noticed that the process of registering posted prices and distributing them was 'cumbersome and slow' in the experimental setting. The process of collecting and distributing prices in the experiment took a long time in relation to a trading period of twelve minutes. If the same procedure had been performed by the Interstate Commission of Commerce in the real bulk freight market, it would not have been as cumbersome, simply because the time taken to collect and distribute prices would have been negligible compared to the period of fifteen days during which prices would have remained fixed. Hong and Plott suggest that the distribution of prices in actual practice might well lead to problems that were absent in their laboratory market, where information was distributed 'immediately, accurately and costlessly', but they provide no evidence for this assertion, and it stands in stark contrast to their own judgement that the posted-prices procedure was tedious.

In a period of fifteen days, negotiators could have walked to the coffee machine many times a day to consult colleagues on the best buying or selling strategy. But there is no time for such leisurely deliberation if telephone lines are constantly busy and any contract at a profitable margin is better than no contract at all. Collecting, distributing and waiting for the price list becomes a burden in those circumstances.

8.11 Scaled-down time and the acceleration of 'data' collection

Plott dislikes comparisons between experiments in economics and experiments in the exact sciences but, in their published article, Hong and Plott wrote that experimenters in the physical sciences 'face the same type of question' as those performing experiments in economics. Let us explore this comparison a bit further by looking at experiments in other sciences. I have chosen two examples: experiments carried out on a scale model of the San Francisco Bay in the 1960s and nineteenth-century experiments on mountain formation. Both are good examples of experiments on physical models that target specific real-world systems, and both considerably compress the time dimension, which is in fact one of the more general virtues of the experimental method.

A few years ago, philosopher of science Michael Weisberg examined a 1960s proposal to turn the San Francisco Bay into a freshwater reservoir. The Army Corps of Engineers was sceptical and, to evaluate the plan, it built an immense scale model in a warehouse, in which the gradient of the banks and the velocity of currents in the Bay, for example, were scaled down so that the effects of different interventions could be measured and compared. The appropriately calibrated scale model was used to evaluate, and ultimately to reject, the proposed project. As for mountain formation, historian and philosopher of science Naomi Oreskes has investigated how, after the Second World War, geologists progressed from nineteenth-century experiments on scale models to computer simulations. The time taken for a mountain to form is of course a good deal longer than a human lifespan. To scale down time, nineteenth-century geologists used different materials, replacing solid rock with clay. They were aiming to find credible causes for visible characteristics of mountain formation, such as fractures and strata in rock.

In both these examples, scientists and engineers had to use alternative materials to preserve features of the real-world system that they were interested in. Copper strips were used to replicate the roughness of the San Francisco Bay and clay was used to replicate solid rock in experiments on mountain formation. Of course the experimental scientists were not interested in the properties of copper strips or clay as such; they chose those materials in order to replicate certain properties of the target system. The copper strips were a suitable choice; clay, as it turned out, was not.

As a final example, take the fruit fly. It does not represent the human being, but with the help of fruit flies the essential features of diseases can be reproduced (up to a point). Fruit flies are smaller and simpler organisms than humans and they reproduce more quickly, so using them in research makes it possible to condense time and to draw conclusions that can then be expanded to fit the human being, based on the notion that the two species share certain characteristics, such as DNA division.

How do examples like these help us to think about a laboratory experiment in economics such as the inland freight transport experiment? In that experiment, both the timescale and the scale of the freight were reduced. One period in the experiment takes four weeks in reality and what constitutes a load depends on the size of the area under study. Hong and Plott claimed that their experiment represented not the real market but only its essential economic features, rather like the materials in our three examples from the natural sciences.

Scaling down trading periods from weeks to minutes makes possible an accelerated collection of data. It also enables the experimental economist to manipulate the market. As an experimenter you can change the rules of price-setting and see what happens; you can use a change in the incentive structure – the incentives given to participants in an experiment – to see how that change will affect market outcomes. You can change market equilibrium by giving everyone a list of prices to act upon (in Figure 8.3, p. 137, those on the demand side were given higher maximum prices in another trading round; the line showing demand has moved outwards) and so on. These different outcomes can then be compared for efficiency, or some other predetermined criterion. Expanding the outcomes of the experiment again to match the scale of the real-world market can render up arguments that help us to evaluate market practices or policy proposals. In this way, an experimenter ensures that he is not just garnering academic knowledge but presenting the means to intervene in the world.

8.12 External validity and the image of markets

Hong and Plott were very much aware that it was not obvious that the data generated in a laboratory could serve to indicate what happens or can happen outside its walls. This is an issue that arises time and again in seminars held by experimental economists in which proposals for experiments are discussed. Before an experiment is performed, its design is explained and illustrated to a group of peers. To an outsider (such as this writer) it is actually impossible to tell from the mathematical equations presented which reality they are supposed to relate to – the person presenting them either assumes without further ado that his or her audience is familiar with such experiments or is so eager to discuss the design of the experiment that the aim of the research (what do I want to show?) receives less attention than it deserves.

Nevertheless, the discussion that follows is always about that aim and about whether the design of the experiment does indeed make that aim visible. It is striking how the participants in such seminars look for examples from the real world that either fit with the host's intentions or show that the design of the research is such that it might perhaps 'stand for' a different reality from the one intended. The extent to which casual reference is made to the role of 'healthy *as if* assumptions' is no less remarkable. The experiment, because of the way it is designed, simulates a situation or event in reality but is not identical with it.

What exactly does this mean? Is the experiment a representation or representative of something in reality? Anyone who stresses the former interpretation will find the problem of external validity far more urgent that anyone who stresses the latter. How, for example, can a single participant in an experiment 'represent' four firms, as in the experiment by Hong and Plott? One participant can of course act as a 'stand-in' or representative of four companies if I, as the experimenter, am interested in ensuring that the aggregated behaviour of the participants in my experiment coincides with the supply or demand function I assign to them. Those functions must resemble the aggregated functions that we, as economists, 'observe' statistically.

Hong and Plott used images as an intermediary between the market in the experiment and the market 'in the wild' – their target market. Look at Figures 8.3 (p. 137) and 8.4. Figure 8.3 shows the market supply and demand schedules that were produced when the supply and demand schedules for individuals were added together (weighted by number of participants). In Figure 8.4, the outcomes of the experiment have been

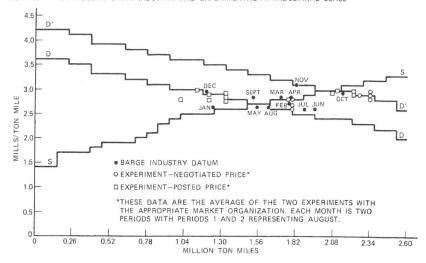

Figure 8.4 Experimental market demand and supply schedules scaled up to the level of the entire market.

Source: James T. Hong and Charles R. Plott (1982). Rate Filing Policies for Inland Water Transportation: An Experimental Approach. *The Bell Journal of Economics*, 13(1), p. 16.

scaled up to the level of the entire market. This is a simple exercise geometrically and arithmetically. But even though Figures 8.3 and 8.4 look the same, they represent very different markets: Figure 8.3 the market in the experiment; Figure 8.4 the 'real' market. Scaling up the market outcomes (visually and arithmetically) changes the nature of the experimental evidence. The large, complex market for bulk freight transport is now represented by them. The numbers in Figure 8.4 are the scaled-up results of the experiment. They are not statistical data gained by making measurements of the full-scale market. That would be impossible, since the pricing policy tested in the experiment was counterfactual. Such diagrams are certainly not the only way of forging connections between the world inside and outside an experiment, but neither should their importance be underestimated, especially when it comes to the practical policy consequences of experimental work.

8.13 Shifting the burden of proof

Hong and Plott's research into the advantages and disadvantages of the posting of prices in advance by freight transport companies demonstrates that creating a market in a laboratory very quickly grew to be of more than merely theoretical importance. Their study was used as evidence to support a proposed policy change by the Interstate Commerce Commission. That change had been requested by rail freight companies in particular, who considered themselves disadvantaged by the prevailing market practices of bulk freight transporters on the inland waterways. The rail freight companies demanded that the water freight companies make their prices known beforehand. At issue was their claim that the market would work more efficiently if prices were posted in advance. The rail freight companies were therefore demanding a change to the usual pricing policy of water freight companies. It is important to note that this research filled a lacuna; no data were available to support the rail companies' argument, since this was all about a different pricing policy from the existing one, in other words it was about a counterfactual.

Hong and Plott's study produced, through experimentation, the data that were lacking, which unfortunately for the rail freight companies pointed to an answer in the negative: a change to the pricing rules would in fact make the market work less efficiently rather than more so. Hong and Plott's experimental study furnished evidence that would otherwise have been unavailable. One argument by the rail freight companies that appeared plausible (if you make your prices widely known then there will be more transparency and the market will therefore work better) was not supported by an experiment that created a market in a laboratory in which that pricing policy was applied.

Theoretical and practical results therefore soon went hand in hand (the same observation can be made in the case of behavioural experiments, incidentally). This generated not just credibility for the results of experiments but an actual market for experiments of this kind. Governments and businesses were all too interested in experiments that might produce evidence to support policy changes or that could be used in lawsuits. Experiments that produce data on counterfactuals can help to shift the burden of proof.

8.14 The experiment in economics

The standard argument, ever since Mill, against the use of the experimental method in economics was that economic phenomena were too complex to allow for controlled interventions. Even though experimental economists such as Plott stressed repeatedly that the laboratory market was simple, it was still a complex thing: consider the number of participants, the many different ways in which they can respond to tasks and so on. The standard argument was not that experiments on individuals were impossible. Mill explicitly wrote about the possibility of carrying out an experiment in one's own mind, since one could be sure of the motives on which one's actions were based. Friedman's argument against the use of questionnaires was that they were 'entirely useless' in testing the validity of economic hypotheses, which to Friedman were hypotheses about how a market economy works. Experiments on individuals teach us nothing about market behaviour; markets cannot be put in a box, so experimental economics is impossible.

The revolution brought about by Vernon Smith, Charles Plott and others was that they made it possible to challenge both these objections. An experiment could be designed in such a way that, in a controlled environment, an economist might learn something from it about the behaviour of markets. The appropriate financial incentives for participants needed to be put in place, trading and communication rules needed to be created and defined with great care, and the experimental market needed to be designed in such a way that it took on the essential characteristics of the target market. In determining what those essential characteristics were, the toolkit developed in the post-war period by econometricians such as Tinbergen was indispensable. Without the statistical characteristics of markets, there would never have been any guidance on how to scale down market characteristics. As for the workings of financial incentives, assumptions had to be made about rational behaviour, but these did not seem particularly onerous, because once economists actually started doing experiments, they saw that the predictions of economic theory that had been difficult or impossible to verify 'in the wild' could easily be verified in the laboratory.

Experimental psychological research, such as that carried out by Kahneman and Tversky, has shown that economists' assumptions of rationality are far from unproblematic, and many of the current controversies in economics, including experimental economics, concern them. But the defence adopted, particularly by economists who perform market experiments, is that the problematic aspects of the rationality assumption have been exaggerated. Market experiments are vandal-proof; even if individuals are not rational in the conditions imposed by the experiment, they act *as if* they are rational.

In this chapter, we have encountered only a few methodological aspects of the experiment in economics. We have looked at the feasibility of replicating markets in a laboratory, at the difference between market experiments and behavioural experiments, at the difference and the connection between simulations and experiments, at issues of scale and – briefly – at the problem of external validity. Based primarily on the study by Hong and Plott, we have investigated how an experiment is designed and implemented and what kinds of conclusions can be drawn from it.

Noticeably absent from the discussion in this chapter has been the currently dominant game theory, a mathematical theory concerning strategic decision-making. Absent too has been a full discussion of the dominant part played by probability theory, especially in behavioural experiments. Of course, both game theory and probability theory generate questions of their own, but these are separate from questions about what an economic experiment is and how such an experiment can be performed. This

148 Experimentation in economics

chapter is intended to throw light on those last two questions, with the aim of making clear how economists have succeeded in rebutting traditional objections to the idea that experiments can be carried out in economics or in turning them around so that those objections work in their favour.

Notes

1 Smith is referring to a study by the renowned experimental psychologist Louis Leon Thurstone, published in 1931, called 'The Indifference Function', in which Thurstone asks a participant in an experiment to compare different bundles of goods. Thurstone settled upon this experiment after discussions with the Chicago economist Henry Schultz. His study is still seen as the first attempt to measure consumer preferences. In what is known as utility theory, an economic agent is seen as someone who tries to maximize his utility, given a limited budget. Another way (and for many reasons a better way) to express this was to imagine a consumer as someone who attempted to achieve as high an indifference curve as he could, given his budget. An indifference curve gave all the combinations of goods that were valued equally highly by a consumer. After the Second World War, it became usual to deploy Samuelson's terminology of 'revealed preferences', in order to avoid any reference to psychological entities such as utility that were impossible to measure (or believed to be impossible to measure).

2 From the early 1940s onwards, Edward Chamberlin carried out experiments in classrooms at Harvard that were designed to test his theory of monopolistic competition. They also served to show the faults in the theory of perfect competition. Until recently, it was thought that Chamberlin tried his experiments for around five years before ceasing to use experimentation in economics. It now seems, however, that Chamberlin carried on experimenting into the 1960s. Aside from his own 1948 article in the *Journal of Political Economy*, a good description of his experiments is lacking, and in later years he seems never to have succeeded in producing a satisfactory analysis of the statistical data he gathered in this way. Whereas Chamberlin called the workings of the market into question, Friedman was one of the most important representatives of the Chicago School, which ever since the Second World War has been one of the most important advocates of faith in the free market. As a student at Harvard in the 1950s, Vernon Smith took Chamberlin's classes. In that same decade, Paul Samuelson performed classroom experiments; although he prepared them for publication, they were never published.

3 Vernon Smith described this problem as one of parallelism between experiment and reality. In the philosophy of science, it is now usually referred to as the problem of external validity, a concept that was introduced into the literature of that discipline, and into economics, as recently as the 1990s.

4 David M. Grether and Charles R. Plott, 'Economic Theory of Choice and the Preference Reversal Phenomenon'. *American Economic Review*, 69(4), 1979, pp. 623–38; Daniel Kahneman and Amos Tversky, 'Prospect Theory: An Analysis of Decision under Risk'. *Econometrica*, XLVII, 1979, pp. 263–91. For a wide variety of reasons, the paper by Kahneman and Tversky rather than the paper by Grether and Plott became a classic and the starting point for the now flourishing field of behavioural economics.

5 In the late 1950s, economist Lawrence Fouraker and psychologist Sidney Siegel joined forces at Penn State University to conduct experiments on bargaining behaviour and group decision-making. That work is still referred to

with admiration by experimental economists as having paved the way for the introduction of the experimental method in economics. Plott is referring here to two different sets of experiments: on bargaining behaviour, in the late 1950s, and on oligopolies, in the early 1960s. Siegel died suddenly in November 1961 at the age of forty-five.
6 As we noted earlier, 'external validity' is not an expression Plott would have used in the 1970s, and in any case he cared little about this problem, in fact it simply did not exist for him. Even laboratory markets were real markets. As far as he was concerned, that was an end to the matter. Vernon Smith's preferred expression for what we would now call 'external validity' was 'parallelism'.
7 For more on this, see Francesco Guala's extremely informative *The Methodology of Experimental Economics*. (Cambridge: Cambridge University Press, 2005, pp. 2–3).
8 For some particularly good examples of these moments when sceptics became convinced, see Svorencik's PhD thesis on the history of experimental economics, *The Experimental Turn in Economics*. (Utrecht: Utrecht University, forthcoming).
9 See for example Smith (Economics in the Laboratory. *The Journal of Economic Perspectives*, 8(1), 1994, pp. 113–31).
10 Plott defends a similar position; see for example Plott (Will Economics Become an Experimental Science? *Southern Economics Journal*, 57, 1991, pp. 901–19).
11 Both Plott and Snow wrote their PhDs along with James Buchanan, a prominent libertarian economist who received the Nobel Prize for his groundbreaking work on public choice theory. Snow received his PhD in the same year as Plott from the University of Virginia. He left the George Washington University Law School in 1975 to become deputy undersecretary at the Department of Transportation.
12 Shortly after they completed their experiments, Hong had a swimming accident which led to a serious brain injury. The paper was therefore written by Plott alone. Plott refers to the accident in the first footnote to the published version of the working paper.
13 For example, the US Department of Transportation's report *The Barge Mixing Rule Problem: A Study of Economic Regulation of Domestic Dry Bulk Commodity Transportation* (1973).

References

Hong, James, T. and Charles R. Plott (1982). Rate Filing Policies for Inland Water Transportation: An Experimental Approach. *The Bell Journal of Economics*, 13(1), pp. 1–19, 1.

Plott, Charles R. and Vernon L. Smith (1978). An Experimental Examination of Two Exchange Institutions. *The Review of Economics Studies*, 45(1), pp. 133–45, 133.

Samuelson, Paul, A. and Nordhaus, William (2005). *Economics* quoted from Francesco Guala *The Methodology of Experimental Economics*. (Cambridge: Cambridge University Press).

Svorencik, Andrej (forthcoming). The Experimental Turn in Economics: A History of Experimental Economics, PhD thesis, Utrecht University.

9 Simulation with models

9.1 Introduction

In Chapter 8, we discovered that it is hard to see experiments and simulation as two separate things, a theme that will recur in this chapter. But we are now approaching the relationship between experiment and simulation from the opposite direction, looking at the place of experimentation in simulation of one specific kind: computer simulation, which uses mathematical, econometric models. Computer simulations are important not just in economics (especially in the context of policy preparation) but in all fields of study that deal with complex systems, such as astronomy and meteorology. In looking at computer simulations we return for a moment to the place of structural models in economics.

Milton Friedman's critique of the work of the Cowles Commission was a serious blow to the builders of structural models, but something known as the Lucas critique looked as if it might be fatal. In 1976, Robert Lucas, Chicago economist and later Nobel Prize winner, published an article in which he showed that parameters in a model's equations could change in response to expectations. Given a strong presumption that expectations will be rational, it is even the case that a future tax rise, for example, will be fully anticipated. The political consequence is that government policy becomes ineffective, while the consequence for macro models is that the structure of the model itself changes in response to policy changes. As a result, the predictions made using such models lose all significance, unless expectations are explicitly modelled within the model. The Lucas critique can therefore be seen as a harsher version of Milton Friedman's.

The critique is valid as far as the predictive power of models is concerned, but the question is whether models are deployed only for that purpose or whether they are used for quite different reasons as well. In Chapter 6, we saw that Carl Christ defends the model of the American economy developed by Lawrence Klein by arguing that it could serve policymaking purposes. Since he formulated his argument in the context of a 'test' of the model for its predictive powers, it was not a strong line of defence even then, but in light of the Lucas critique it seems completely untenable. Prediction and policy are inseparably linked in this critique. But what if the primary purpose of macro-economic models is not to make predictions but rather to determine possible courses of action in response to an assessment of a concrete socio-economic state of affairs? For policymakers, models may indeed be useful mainly as an aid in carving out such strategies. In the Dutch context, where the predictions made by the CPB actually have significance in law (the government is legally obliged to take the CPB's predictions as its starting point when putting together a budget), this does not seem credible. Yet it is.

Let us briefly return to the first ever macro-economic model, the model of the Dutch economy presented by Jan Tinbergen in his 1936 preliminary advice to the (now Koninklijke, Royal) Vereeniging voor de Staathuishoudkunde, of which the secretariat has been based since time immemorial at the Central Bank of the Netherlands. At that meeting, it quickly became clear that the members were unaccustomed to receiving advice in a mathematical form. Jan Goudriaan, who had trained as an engineer at the Delft University of Technology and was therefore more than capable of reading and understanding Tinbergen's model, quoted his old mathematics teacher in Delft, who compared systems of algebraic equations with the 'night train' as opposed to the 'day train' of geometry. The comparison went down well with those present, many of whom had the impression that the 'mathematical machine' (as Tinbergen called his model) was a 'night train' from which conclusions rolled out that they would simply have to accept. His work and efforts were praised but there was discomfort and unease about the form of his arguments.

This applied in particular to the crossing of boundaries that Tinbergen had permitted himself in the field of exchange-rate policy. In 1936, the Netherlands was the only country still sticking to the gold standard and opinions on that point among politicians, economists and statisticians were sensitive and divided. Economists had been asked to stay away from the subject but Tinbergen paid little heed to that request. His preliminary report consisted of what he referred to as 'railway timetables'. These were a range of possible policy scenarios, calculated for their effects with the help of his model. His most important advice was: drop the gold standard. This had already happened elsewhere. In practice, it came down to a devaluation of the guilder but, without proper insight into the workings of the model, it was difficult for those present to understand the reasoning behind this politically sensitive scenario, even if they supported the abolition of the gold standard for other or indeed for the same reasons. Tinbergen did his best to explain how his model worked, hoping to turn the 'night train' into a 'day train'.

Such a scenario is nothing other than a simulation.[1] Changing something in the model creates a 'what if' story. We say of a model that it simulates reality if, based on certain criteria, it resembles something in the real world, but with a 'what if' story we are talking about simulation in a more far-reaching sense. Such stories, after all, are not about something that resembles reality, since they deal with things that have not, or have not yet, taken place. Tinbergen used his model to simulate what would happen to the Dutch economy if the Netherlands abandoned the gold standard. There is a fundamental difference between this and a prediction in which the values of variables for the future are based on extrapolations from the present.

An intervention in a model, such as leaving the gold standard, is also known as an experiment on a model. It is a virtual experiment – to use a useful term coined recently by historian and philosopher of social science Mary Morgan – in the sense that there is no actual intervention in reality but only an intervention in the reality of the model. It seems, however, that in this latter case a model can tell a credible story – simulate the economy credibly – only if it credibly resembles the reality in the first sense (for example the actual Dutch economy). We have to believe that the model is a credible representation of reality before we carry out any useful 'experiment' on it. Note the difference between this and Samuelson's thought experiment. His model was constructed to test the credibility of an important intuition held by economists (and to undermine that intuition), whereas in the case of Tinbergen's model, its credibility had to be assured (it had to be accepted as a credible representation of the Dutch economy) before a policy change was implemented in the model.

But on what grounds can we say that a model resembles reality and what does 'reality' mean here in the first place? According to Friedman, the criterion lies in the accuracy of the model's predictions, but in the case of simulation that criterion does not apply. So to judge the quality of a simulation, criteria other than accuracy of prediction are needed.

In this chapter, we shall look at those other criteria, taking as our example the econometric model developed by the Central Bank of the Netherlands in the 1970s, which was used until about 2005 (if in an altered form) in the preparation of its monetary policy. The model is called MORKMON, an acronym that stands for MOnetair en Reëel Kwartaal MOdel van Nederland (Monetary and Real Quarterly Model of the Netherlands). We will discuss the development of this model against the background of a test of an extended version of the macroeconomic model constructed by Lawrence Klein that does not refer to its predictive power. This test was first proposed, described and used by the Adelmans, a married couple, so it became known as the 'Adelman–Adelman test', even though the Adelmans actually carried out three tests. This background helps us to think about the significance of model simulations in the economy and the role of economic expertise, especially in the context of policymaking.

To be able to see how the introduction of model simulations changed preparations of monetary policy, we first need to look at how monetary policy was prepared in earlier years. From the early post-war period, we now move on to the late 1960s, to the research department of the Central Bank of the Netherlands, where the man who would later become its president-director and subsequently the first president of the European Central Bank, Wim Duisenberg, was appointed special advisor to the directors.

9.2 The research department of the Central Bank of the Netherlands

The first post-war president of the Central Bank of the Netherlands, Marius Holtrop, took up his post in 1947 and stepped down in 1967. Holtrop was an authoritarian but popular leader who had written his PhD thesis on a monetary subject in the interwar period (the rate of circulation of money) before taking a series of management positions in the business world (at Shell and at the Dutch steel company Hoogovens). He emerged from the war with his reputation intact.

Immediately on taking up the presidency of the Central Bank in 1947, Holtrop appointed a number of heavyweights to the research department, realizing that the bank's authority would depend to a great degree on the quality of its analysis (Figure 9.1). Along with the head of the research department, G.A. Kessler, Holtrop developed what became known as the bank's 'normative impulse analysis', a diagnostic instrument for monetary policy. The aim of monetary policy was to keep the liquidity quotient (the relationship between the money supply – as measured by M2 – and the nominal national income) constant, so as to ensure that the circulation of money in the economy was at the right level, which is to say 'neutral'. Holtrop and the research department then traced where disturbances ('impulses') came from (the public, abroad, the government, the commercial banks) and whether they could be controlled or steered by the Central Bank.

The statistical work of the research department, the official source of statistics in the monetary field, was extremely important to those carrying out this analysis. The analysis itself was laid out in the Central Bank's annual report, of which Holtrop always wrote the initial version before having the document passed back and forth between him and the research department (especially Kessler) from November to late March. The monetary analysis they presented amounted to a legitimization of the policy implemented over the previous year, while at the same time it offered a preview of the policy that could be expected in the near future from the Central Bank.

Holtrop's monetary analysis came to be known as 'Dutch monetarism' and it found a place alongside Milton Friedman's better-known and far more influential monetarism. Like Friedman, Holtrop had a low opinion of the then popular Keynesianism, which placed the emphasis on government spending policy.

154 *Simulation with models*

Figure 9.1 Directors of the Central Bank of the Netherlands in 1958. Seated from left to right: A.M. de Jong and H.R. van Talingen. Standing from left to right: A. Houwink, S. Posthuma, M.W. Holtrop and J.H.O. Graaf van de Bosch.

Source: Reproduced with the permission of the Central Bank.

Contemporaries described the approach of Holtrop and the research department as 'detective work'. From the available statistical material and from 'circumstantial evidence', a picture was built up of the monetary events of the year in question and finally 'culprits' were sought, in the form of factors that had caused inflationary or deflationary pressure on the money supply. The Dutch Central Bank's monetary analysis was therefore strongly bound up with Holtrop's personality and with the interaction between Holtrop and the research department. Special talents were required for this kind of detective work. When Holtrop left in 1967, the analysis lost its driving force and organizational brain. Those working at the research department at the time recall that the monetary analysis in the annual report became increasingly rudderless, with departmental staff adding pieces of text that had earlier been removed, only to take them out again. This would go on until the absolutely final deadline for the report arrived.

As a result, the Central Bank lost a great deal of its clout with the various organizations that dealt with socio-economic policy in the Dutch economy (an economy characterized by consultation and consensus).

Holtrop had never had much respect for the CPB's 'silly behavioural equations', but acceptance of his findings had always depended on the authority he bore. He personified monetary analysis and, after his departure, the Central Bank started to lose out, in the public policy realm, to the CPB analyses based on the econometric modelling tradition that had developed in the wake of Tinbergen's work. Under their new president Jelle Zijlstra, the Central Bank's directors therefore felt the research department needed to change.

Wim Duisenberg became special advisor to the board of directors at the Central Bank of the Netherlands in 1969. After finishing his PhD thesis, he spent a number of years working in the research department at the IMF under J.J. Polak. In 1939, Polak had followed Tinbergen to the League of Nations, where he worked on Tinbergen's research into business-cycle theory. When the Second World War broke out, Polak left for the United States, where he joined the IMF and became extremely influential as head of its research department. It was around then that large-scale econometric modelling became the standard means of monetary analysis and policy preparation in places including Italy, Canada and the United States. Evidence for this includes the econometric model that was developed at MIT and the University of Pennsylvania by Franco Modigliani (along with quite a few graduate students at MIT and the University of Pennsylvania), which explicitly attempted to make an integrated model of both the monetary side of the economy and the real, productive economy. Duisenberg advised the Dutch Central Bank to develop its own econometric model, so that it could keep up with what was happening internationally. This meant it would be necessary to set up an econometric research department within the Central Bank. By developing its own integrated model, the bank hoped to regain its policy clout vis-à-vis the CPB.

This new department was led by econometrician Martin Fase, who had studied at the University of Wisconsin, an important centre for econometric research in the mid-1960s. The econometric research group at the Central Bank was soon made up of six people. Most had been brought in from outside; a few made the transition from the old research department to the new group. In the 1970s, the new econometric research group found itself in an uncomfortable position in relation to the old research department. It was only when a separate department for Scientific Research and Econometrics was set up in 1983 that this animosity ceased – for a while, at least.

The development of the Central Bank's integrated 'monetary and real model' took some ten years altogether, because much of the preparatory work had to be done from scratch. This is often forgotten; in those days the development of a structural model was an extremely labour-intensive and time-consuming affair. Making a model was akin to setting up an engineering project. There were several phases. First, based on a new CPB model, a quarterly model was constructed by the new econometrics research group at the Bank as a way of gaining experience. Then partial studies were made of (among other things) equations that would inevitably be part of the complete model, such as those concerning bank credit, the mortgage market, foreign capital traffic and portfolio demand from the private sector. The basis for the complete model was a system of financial accounts, a coherent system of balances that had been developed under Holtrop's presidency and was at the basis of monetary policy analysis during his time in office. An important decision that determined the direction of the entire model was: for which balance sheet would behavioural equations be generated and which item would ultimately be treated as the balancing item? Implicitly (and the originators of the model were very much aware of this) such a decision implies opting in favour of a specific 'moment of action'; it defines the causal structure of an economy. Once that and other choices had been made, software had to be developed to perform all the model's calculations, datasets had to be collected and made consistent with each other and so on.

Figure 9.2 shows the steps that needed to be taken in constructing the model. It demonstrates that the development of an econometric model not only took a great deal of time, it really did look like a complex engineering project. For the econometricians who were involved in the project throughout, the final phase, in which computer simulations were carried out with the model, was of course the most interesting.

This short history of the research department of the Central Bank of the Netherlands indicates that its development of an econometric model served a clear politico-strategic aim: the reinforcement of the bank's position in policy preparation. This was achieved by matching the CPB's detailed level of argumentation and by re-establishing the Bank's international credentials in monetary research. It also shows that the development of the model led to changes in the internal organization of the research department. In place of the personal judgement of the Central Bank's leadership, the impersonal model came to the fore. This would not only make the bank's verdict on the state and development of the monetary economy objective, which is to say independent of its president's judgement, it would refine it as well. As with the business-cycle research at the CBS that we looked at earlier, this presumed a new kind of organization. A model is not just a system of equations that are tested against statistical data, it requires an organizational and administrative infrastructure. The development schedule of the model shows rather neatly that by the time the task is complete and the model is 'ready' for use, an organization will have been built that can maintain the model. But as the remainder of this chapter will demonstrate, no model can establish its objectivity without economic expertise being built into it.

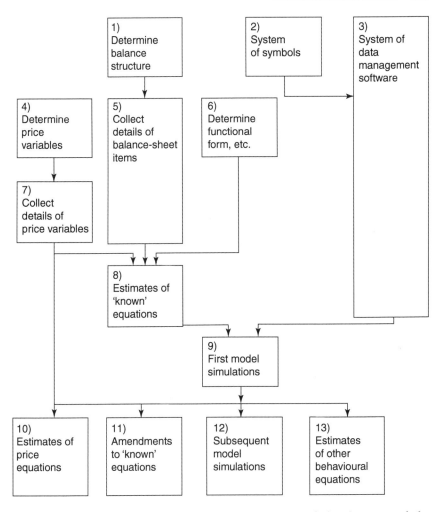

Figure 9.2 Diagrammatic representation of the stages of development of the monetary model of the Central Bank of the Netherlands.

Source: The personal archive of Frank den Butter; internal memorandum from F.A.G. den Butter to Professor Fase, 'MOMO; reaction and proposals with reference to your schedule of balances', 3 April 1979, p. 6. Courtesy of Frank den Butter.

9.3 Model simulations: the 'Adelman–Adelman test'

Holtrop did not just talk in an offhand way of the CPB's 'silly behavioural equations', he also stressed time and again that CPB models concentrated purely on the short term, whereas his monetary analysis related to the 'fundamentals' of the Dutch economy in the long term. In saying this he was not merely doing the CPB an injustice, he was failing to acknowledge that the work of the CPB consisted only in part of making short-term predictions, for the benefit of the national budget for example. Following on from Tinbergen's 'railway timetables', the CPB also developed 'scenarios' whereby possible paths along which the Dutch economy might develop were laid side by side and then coupled with urgent issues (think of today's concerns about the aging of the population, for instance, or the need to develop sustainable sources of energy), usually in combination with various policy options available to the government. One important part of the CPB's work, in other words, was to carry out this kind of scenario analysis – to simulate different trajectories of the economy based on different assumptions about policy measures or international developments, such as changes to international exchange rates or fluctuations in the growth of world trade. It was precisely these simulations that gave the CPB the edge over the Central Bank in dealing with national institutions that are involved in policy preparation.

How did the Central Bank of the Netherlands develop its own model simulations? To answer that question, I first want to look back at a 'test' carried out by Irma and Frank Adelman in 1959 on an adjusted version of the macro-economic model built by Lawrence Klein (the Klein–Goldberger model). In their much-discussed 1959 article in *Econometrica*, the Adelmans did not ask themselves whether the Klein–Goldberger model produced good predictions for the American economy but whether it offered a good representation of it. To predict and to represent are two very different things. With Friedman's 'as if' criterion you engage in the former but never the latter. What the Adelmans were interested in was the reverse: whether the model gave a good representation of American business cycles rather than whether it rendered up accurate predictions. Predictions are almost always off-target. We simply cannot expect accurate predictions from economic models but we can expect models to capture important characteristics of complex real-world systems. The Adelmans' criterion for a good representation was simple: if the model could produce a statistical pattern over a long period that an expert could not distinguish from available statistical data patterns (such as those provided by one of the most important statistical bureaus in the United States, the NBER), then it was a 'good' model.

Three steps were taken to arrive at this test. First, they entered initial values for the American economy for a specific year into the Klein–Goldberger model and calculated the behaviour of the model for a period of a hundred years. Did the model simulate, in other words represent, the American economy? No, as it turned out. Even though a linearized model was capable of showing a cyclical pattern, this model did not (see Figure 9.3). As a second step, the Adelmans gave the Klein–Goldberger model a 'kick', which is to say that they made a one-off change to one of the exogenous variables (in their case a one-off radical reduction in government expenditure). A considerable change like this knocked the model out of balance, but after a number of 'years' it returned to the original values. So the model was still not showing the desired pattern in the data. There was still no cyclical movement.

The third step the Adelmans took was to add 'white noise' or tiny shocks that were determined by chance. The effect of this third treatment was miraculous. Instead of a linear progression, or those major swerves caused by any dramatic change in one of the initial values, the model now showed a cyclical pattern that (given a certain degree of leeway) was indistinguishable 'with the naked eye' from the actual course of the American economy (see Figure 9.4). To the Adelmans, 'with the naked eye' meant that the development of the individual variables in the model corresponded with actual, observed developments in the American economy. But they added an important qualification with regard to whose naked eye would be able to observe the cyclical pattern properly: 'If a business cycle analyst were asked whether

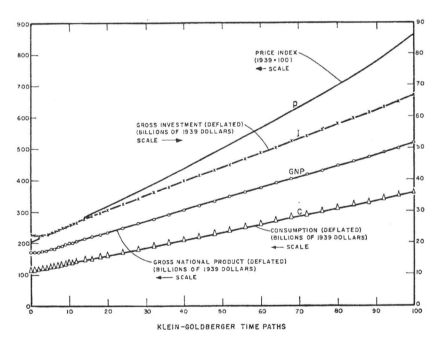

Figure 9.3 Graph of the Adelmans' simulation of the Klein–Goldberger model without the application of a 'shock'.

Source: Irma Adelman and Frank L. Adelman (1959). The Dynamic Properties of the Klein–Goldberger Model, *Econometrica*, 27(4), p. 601.

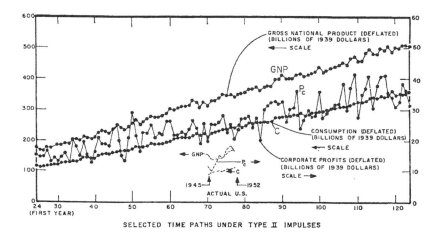

Figure 9.4 Graph of the Adelmans' simulation of the Klein–Goldberger model after the application of random shocks.

Source: Irma Adelman and Frank L. Adelman (1959). The Dynamic Properties of the Klein–Goldberger Model, *Econometrica*, 27(4), p. 608.

or not the results of a shocked Klein–Goldberger computation could reasonably represent a United-States-type economy, how would he respond?' (Adelman and Adelman, 'The Dynamic Properties of the Klein–Goldberger Model', *Econometrica*, 1959, p. 612). It was not just anyone but a 'business cycle analyst', someone with intimate and lengthy experience of economic statistics, who must be unable to detect any difference – in which case the model was 'good'. Dutch philosopher of economics Marcel Boumans has compared this qualification with the Turing test for artificial intelligence. If a person is unable to distinguish between the 'output' of a computer and that of a human being, then the computer has simulated human intelligence. A similar criterion was formulated by the Adelmans for economic models, except that the 'someone' would have to be an expert on economics; he would have to know what a business cycle looked like.[2]

In the case of the Klein–Goldberger model, such an expert would, for example, look at the overall pattern of a number of variables over time. It would be strange if national income increased while private consumption and business investment were falling sharply, just as it would be odd if a business cycle had very high peaks and exceptionally deep troughs. The Klein–Goldberger model scored well on all such points, so the Adelmans concluded that the model was 'not very far wrong' as a representation of the American economy.

9.4 Model simulations at the research department

One of the main collaborators at the Central Bank had actively examined the Adelman–Adelman test for his Master's thesis. Nevertheless, although the model simulations of the econometric study group followed the Adelmans in some ways, in other ways they did not. As with the Adelman–Adelman test, it was important that the model offered a good representation of the characteristics of, in their case, the Dutch economy. But it was also important that the model could be used for purposes of policy evaluation, so characteristics of economic statistics such as the amplitude or length of the business cycle were built in 'by hand', not by adding noise (although that was added as well). Building the different elements of the model therefore came close to 'curve fitting': the model's equations were adjusted to fit the actually observed course of the variables, very much in the spirit of Tinbergen. Then, as with the Adelmans, 'shock tests' were carried out on the model. The main aim of these tests – and in this sense the Central Bank econometricians diverged from the Adelmans – was to gain insight into the structure and workings of the model. Through which channels was a shock transmitted? Did the model perhaps consist of different, relatively separate 'blocks' that could be studied independently of each other? Whereas the Adelmans had in a sense treated the model as a 'black box' and ultimately judged its quality purely based on its output – the statistical data that rolled out of the simulation – the Central Bank econometricians, more in the tradition of Tinbergen and the Cowles Commission, looked at the internal structure and coherence of the model. On that basis it was possible to tell a story of causality, of the way in which changes affected the economy. Applying shocks to the model therefore served not only to highlight its weak points but to turn a 'night train' into a 'day train' by bringing out the model's causal structure.

In the context of Dutch public policy, this made sense. Institutions such as the CPB and the Central Bank of the Netherlands were officially and unofficially expected to give advice on proposed government policy or on the implications of a sudden change in the economy. In the late 1960s, it was precisely in this respect that the Central Bank lost out to 'the number guys at the CPB', who, armed with their models, were able to tell more coherent, detailed and verifiable stories about the economy than the Central Bank's research department, which relied on 'detective stories' that were increasingly hard to explain to the outside world. The Central Bank econometricians were now trying to develop such stories with their own model.

More coherent: the equations in the model could of course not contradict each other, but they also had to meet a less strict condition, in that they could not operate upon each other in such a way that a 'less than credible picture' of the economy emerged. As with the Adelmans, expert knowledge of an economy was important in judging whether the model was working well or not.

162 *Simulation with models*

More detailed: by tracing the results of a shock to the economy-in-the-model, it became easier to see the channels through which a shock was transmitted. With the 'detective work' of Holtrop's impulse analysis, this level of detail had been unachievable. The model was also more detailed because the search for a cause of inflationary or deflationary pressure did not have to be limited to the diagnosis that it lay 'with the banking sector'. A far more detailed account might now emerge, showing that within the banking sector a number of variables could operate upon each other and suggesting how they did so.

More verifiable: when the staff at the Central Bank of the Netherlands tried to replicate the quarterly model that the CPB had just developed, they encountered all kinds of difficulties. First of all, it was not the usual practice at that time to publish the data, nor indeed all the specifications of the model. This meant that for outsiders it was practically impossible to reproduce the results the CPB had derived from estimates and simulations. Although such an exercise remains extraordinarily difficult to this day, model specifications and datasets are now usually made available to third parties. There is another reason why verifiability was an important motivation for the Central Bank to develop a model of its own. The Central Bank's monetary analysis had, up to then, been highly dependent on the authoritative Holtrop. A verifiable model would in and of itself – at least, such was the reasoning – confer 'authority'.

In the Dutch context, then, much attention was paid to the inner coherence and structure of the model. Model simulations using single shocks helped in investigating this inner coherence, but applying a single shock to the economy was also very much in keeping with the task of evaluating the implications of policy. The Central Bank staff 'practised' by calculating the effects of pay claims made by trade unions, the implications of election manifestos or the results of proposed taxation measures. Quantitative assessments such as these had to produce credible outcomes – but what did that mean? How is it possible to tell whether a model is producing credible, realistic outcomes for situations that do not yet, indeed may never, exist? Before we look more closely at how the econometric study group answered this question, it may be helpful to consider a few points of similarity with the example given in Chapter 8: Hong and Plott's experimental study of price-setting in freight transport by water.

9.5 Scaling down an economy

We have seen that with their experiment, Hong and Plott claimed they had overcome a number of barriers in statistical, econometric research. With the help of econometrics, it was possible to make estimates concerning the structure and dynamics of a market such as inland water transportation, including estimates of supply and demand elasticity, but not to intervene materially in that market. A statistician or econometrician could make an estimate only based on the past. At best, he could predict the future assuming the economic structure remained unchanged (an assumption that, over a long period, is simply untenable).

Hong and Plott's scaled-down market enabled them to intervene materially; they could change supply and demand schedules, alter the price-setting mechanism and so on and examine the results. This meant they were in a position to produce data that would otherwise simply have been unavailable. They could condense time and so produce more data than they could have done based on 'real time' or time outside the laboratory. Of course, faith in these data (on the part of the judiciary, the government or the general public) depended, as it still does, on the potential to scale up the resulting data to the level of the 'real' economy without encountering problems.

In fact, something similar applies to the use of models for simulation. The Adelmans calculated the Klein–Goldberger model for a period of a hundred years. They did so on a state-of-the-art computer, which in those days meant an IBM 650, with a computing power roughly equal to that of the dashboard of a modern car. The Adelmans could calculate the results for one year in one minute. With their simulation they were therefore in a position to produce data in a hundred minutes that had taken the real American economy a century. These were fictional outcomes that derived their credibility from a comparison with the time series available for a period of the same length in the American economy. Just as the study of Hong and Plott scaled up experimental trading rounds of minutes to weeks, time in the Klein–Goldberger model was scaled up from minutes to years.

In an early simulation exercise by one of the Central Bank's econometricians, the time taken to calculate the figure for one year was reduced to 0.17 seconds. This was a simulation carried out in the early 1970s using a Dutch-built ELX8 computer, which was so fast that the CPB used it for its own calculations, sending couriers from The Hague to Amsterdam with fat parcels of punch cards for the purpose. The lesson from all this is that economists can place themselves 'outside time' not only by means of experiments but also by performing simulations. They can condense time at will and by doing so create data that would not otherwise exist.

According to the criterion used by the Adelmans, such fictional data were 'good' because an expert economist would be unable to see any relevant difference between the simulated figures and the actual figures. They were 'good' because the expert looking at the figures could not tell anything about the period in which they were generated. Imagine yourself working through time-series charts from the past, covering up some of the information available and simulating an extrapolation as a way of testing your expertise at reading time-series data. This is, in fact, precisely what some of the greatest economists have done over the years. Reflecting on a paper by Kathryn Domingues, Ray Fair and Matthew Shapiro,

'Forecasting the Depression: Harvard versus Yale', published 1988 in the *American Economic Review*, Paul Samuelson reflected:

> Many years ago I used to amuse myself by shuffling through past time series charts, while covering up knowledge of the data involved and revealing to myself only gradually how the future turned out. Often my simulated extrapolations were near the mark; sometimes, though, not ... I discovered that I would have been wiped out in Fall 1929. There was really nothing terribly distorted about the previous long epoch of equity prices.[3]

The most successful test by the Adelmans was the addition of statistical 'white noise' to the model. I have already indicated that this was not the route taken by the Central Bank or the CPB in the 1970s. What they did instead was to give the model a shock, which could be seen as an intervention in the model or as an experiment with the model. How do you know whether such an intervention is 'good' or 'successful'? That is the same question as was posed in section 9.4: how can you tell whether a model outcome that does not yet exist in reality is 'realistic'? One obvious answer is: wait and see whether the outcomes do actually come about. But many such interventions are never made *in vivo*, only in the model. By experimenting on the model, therefore, you can generate data such that it appears as if the future is already present. Such data seem like predictions but that is not what they are. They are better seen as forecasts; they show what can happen rather than what will happen. After all, no one is saying that an intervention in the model reflects something that will actually be carried out in reality, let alone in all conceivable cases. The quality of this kind of experiment on a model cannot therefore be simply read from a criterion such as the precision of its predictions. Its quality has to be judged based on other criteria. What those criteria look like became one of the most important questions the Central Bank econometricians faced but, before we move on to that, we need to consider an issue that has been lurking between the lines throughout this chapter: the difference between model simulations, or experiments on models, and laboratory experiments.

9.6 Simulations, model experiments and laboratory experiments

From a brief comparison of Hong and Plott's experimental research with model simulations, it almost seems as if there is no genuine difference between an actual, material experiment and an experiment on a model. Both compress the time dimension, which is one reason why they both enable the investigation of counterfactuals. Both are in need of a benchmark that bears a resemblance to the real world. In material experiments, this latter condition is not generally necessary. Indeed, the ability to create markets from scratch proved to be one of the factors that led to the rise of the experiment in economics. Experiments were then used as test beds, as ways of making controlled changes in trading rules, for example, so that different experimental outcomes could be compared based on a pre-selected criterion such as market efficiency. In the next step such experimentally created 'synthetic' markets were implemented in real-world conditions. A rather apt comparison can be made with the patient information leaflet included with all pharmaceuticals. Implementation of an experimental market in the real world involves the real-world creation of laboratory conditions: take this medicine under the following conditions and it will produce the following effects.

Economists tend to use the term *ceteris paribus* for such conditions, usually translated as 'all other things being equal'. In comparing model simulations (or experiments on models) with laboratory experiments, it may be more useful to stress issues of control rather than the issue of whether 'all other things' (whatever they are and wherever they come from) remain the same. Issues of control differ significantly between the two types of experiment. In an economic laboratory experiment, experimental economists try hard to control for all circumstances that fall outside the specific question the experiment is designed to answer.

For example, as we have seen, the instructions should appear neutral so as not to predispose the participants to act in a specific way. The incentives for participants should be designed such that the experimental economist can have confidence that what they do will be in line with the design of the experiment. The experiment should be carefully explained, so that all participants understand what they are supposed to do (read prices from left to right or from top to bottom; buy rather than sell and so on). In an experiment, causal factors that might otherwise influence behaviour should be excluded, and those conducting the experiment will make every effort to rule them out or to control for them through the way their experiment is designed (examples might include gender issues, differences between students of economics and other students or the effects of the euro crisis). The only remaining issue must be that which the experiment is intended to address.

In our earlier example, this will be: which of two pricing rules comes out looking best in terms of market efficiency? It is perhaps important to note that those conducting the experiment may well have an idea about which of the two rules should emerge the winner but what happens must depend on what the participants actually do. Although the experimenter will define the boundaries within which the participants act, those actions must not follow mechanically from the boundaries set. The prevailing image we have of an experiment is of a scientist controlling all the conditions in the laboratory so that the effects of making one single controlled variation can be measured. To some extent, this is mimicked in the study

166 *Simulation with models*

by Hong and Plott: impose maximum control, provide participants with supply and demand schedules, change the trading rules and then measure what happens. By shielding the laboratory from the world outside, an open space is created within four walls in which observations can be made.

It is a different matter with model simulations or with experiments on a model. Here, no attempt is made to control for external conditions that influence economic outcomes. In the model simulation carried out by the Dutch Central Bank, foreign interest rates clearly do influence economic outcomes but no attempt is made to control the foreign interest rate in a particular modelling exercise, and for good reasons: there is no causal link between foreign interest rates outside the walls of the Dutch Central Bank's research department and the model the econometricians are working on. There will be a stand-in for a foreign interest rate within the model, but that variable is not the actual foreign interest rate. Whereas in a scientist's laboratory, deliberate action has to be taken to shield experiments from the external world, in model simulations (or experiments on models) no such action is necessary. Once the model has been specified and the necessary software is in place to enable numerical calculations with it, the model will churn out results regardless of what takes place in the outside world. Moreover, the model will produce those results mechanically, which is to say without the involvement of any agency of any kind (given that it would be misleading to refer to a probabilistic process added to the model as an agency).

In model simulations, various possible states of the world can be fed into the model to see how it responds. This is exactly what took place at the Dutch Central Bank's econometric research department. Some of those possible states were based on real-world situations; indeed the department tried to get as close as possible to those real situations by mimicking their statistical characteristics, and circumstances in the outside world served in turn to determine a benchmark for the quality of the model. In the next section, we look in more detail at how this works.

9.7 Basic prognosis and intervention

How can policy that has not yet been put into effect, or developments in the world economy that have yet to take place, be evaluated? That question seems to contain its own answer. A situation of unchanged policy, or an unchanged world economy, can serve as the basis for comparison against which a model simulation, or an experiment on a model, can be set that includes specific policy measures. In other words: in the first case you make an extrapolation from the present day, while in the second case you change the parameters (or specifications) in the model. You then calculate results for both and compare them.

If only it were that simple. What initial values should be used for calculations in the model without policy changes? Simply the values of the present moment? For an open economy such as that of the Netherlands, this is far from self-evident. After all, we know that economic development in the Netherlands is highly dependent on factors such as the exchange rate of the dollar against the euro, the price of oil and developments in world trade, to name but a few. There is a fair chance that simply making a quantitative assessment on the basis of continuity will produce an unrealistic picture of the future. So to achieve a 'good' picture, we would do far better to take account of knowledge about the economy that comes from outside the model. Expert knowledge offered by economists must in some way be built into the initial values. This knowledge must be included in the process of choosing the initial values on which calculations will be based. The Central Bank econometricians came to the conclusion that the initial values of a model experiment needed to be 'plausible'.

On the one hand, this raises the question of whether the kind of knowledge that an expert has of an economy outside the model, which enables him or her to arrive at these plausible values, might be of the same character as the knowledge Holtrop had built up over years of experience, years in which he shuffled past time-series charts to get a 'feel' for their movements. On the other hand, it means that knowledge based on a 'plausible judgement' but not on an actual state of affairs is, as it were, hidden away in the model; what looks like a neutral projection of the model – an extrapolation from the present – in fact contains a judgement, made by a group of expert econometricians, that is invisible to the public. A seemingly transparent and objective comparison of a simulation with and without a policy change is not independent of a judgement by a group that determines what is a 'plausible' reference point and what is not. This judgement ultimately rests on a consensus of experts, not on 'the facts', because the facts that are needed to make the base projection and the intervention are simply not there, and perhaps never will be.

Does this matter? There is certainly an issue here that is not limited to simulations that economists make by means of models. To take a relatively recent example, the plume from the Icelandic volcano Eyjafjallajökull disrupted air traffic in April 2010 on the basis not of actual observations of ash particles but of model simulations of the spread of the ash cloud across Europe. After a few days, airline companies naturally started to protest at a flight ban that was, according to the president of the German airline Air Berlin, 'based purely on computer simulations'. In the context of problem-free test flights by KLM and Lufthansa, he seemed to be implying that such simulations did not reflect reality. Damage to a number of F-18 and F-16 jet fighters that flew through the ash cloud showed that the situation was rather less simple.

The attention of anyone looking for recent economic examples will no doubt be drawn to the valuation models of the complex financial products involved in the financial crisis of 2008, or to the 2010 'stress tests' for banks

168 Simulation with models

that were presented in evening news programmes as if their results were hard facts. These models too are based largely on simulations. Anyone dismissing simulations because they incorporate only plausible values (such as a plausible estimate of the likelihood of a financial crisis occurring) is forgetting that simulations, or experiments on models, give us something we would not otherwise have: insight into a reality that does not yet exist, which enables us to form a judgement as to whether that reality is desirable or, indeed, plausible.

This means that the combination of the judgement of economic experts and model computations is essential. Nothing less will do. In a syllabus about modelling used by the CPB for educational purposes, this interweaving of model and experts is expressed in a simple flow chart (see Figure 9.5). It clearly shows that parameters in the model are not merely estimated; they can be, and indeed are, adjusted by experts. The same goes for the initial values for data used in simulations.

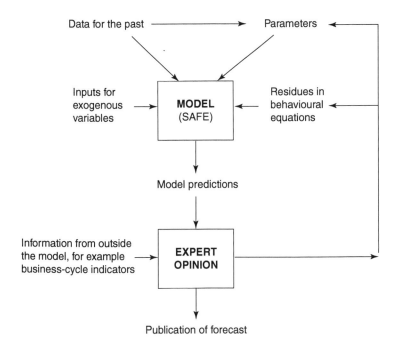

Figure 9.5 Diagram showing the relationship, as seen by the CPB, between expert knowledge and the model. The diagram relates to short-term forecasts and analyses (for which the SAFE quarterly model was developed by the CPB in the mid-1990s), but it is valid generally for CPB models and a fortiori for the use of models for scenario analysis.

Source: The syllabus for the BoFEB course (a course in economics and finance for policy-makers in the public sector), 29 October 2009, Johan Verbruggen, CPB. Courtesy of Johan Verbruggen and CPB.

In short, it seems we have to choose between two evils. The government or the public can place its trust in the judgement of someone it recognizes as an authority, or it can place its trust in a similar judgement that is at least subjected to the discipline of the model. In the latter case, the judgement of an expert has to comply with the criteria imposed by the model: coherence, detail and verifiability. In neither case can we escape the subjectivity that is an inevitable part of an expert judgement.

9.8 The 'night train' and the 'day train'

In this chapter, what are known as structural models have been resurrected, not so much as predictive models, even though they do have an important and legally defined task in that sense in the Dutch context, but as models that enable us to carry out simulations. Simulations vary in character. In our example of the structural model developed in the 1970s by the Central Bank of the Netherlands, the simulations could be described as experimentation on a model. With such experiments it is possible to collect 'data' that give us information about a fictional course of events. This does not simply turn a model simulation into fiction, since it gives us information we could not otherwise acquire, or only with difficulty, concerning a situation about which we are seeking answers. Of course, such an experiment may also generate new questions but, in the context of policy at least, the stress tends to lie on the answers rather than on further far-reaching research. The outcomes of a model simulation are not predictions, they are forecasts. The one-sided emphasis laid on the accuracy of predictions by those who adhere to the line taken by Friedman is therefore misplaced.

In model simulations, it is important that insight is gained into possible mechanisms present in an economy. In that sense, a model simulation is not carried out purely to present results but also to show how those results were achieved. Tinbergen's 'railway timetables' show the route taken, among other things, so a simulation experiment is intended to be not a 'night train' but a 'day train'. Structural models are often so complex in character that even the researchers are surprised when they see how the model responds to a shock. It is the analysis they make in retrospect that shows what route was taken.

The value of simulation experiments stands or falls by the faith we can have in the underlying model. The Adelmans formulated the criterion that an expert must be unable to detect any difference between a simulation using the model and real datasets. This criterion is used to this day as a benchmark for a 'good' model. The idea is that on this basis we can have faith in experiments that are made using the model but it remains a question of trust. This applies not just to the so-called soft sciences but to models developed in the hard sciences as well, including climate models or the simulation models that are used to study the origins of the universe. Economists like to present their model computations as objective, based on the fact that they are quantified, but objectivity in the realm of a model science such as economics is a chimera and faith in numbers masks the omnipresence of the judgemental expert.

Notes

1 Not worked out in those days with an electronic computer, of course, but by the human computers of the CBS.
2 It will come as no surprise to learn that the same criterion applies for models in a broad range of sciences, from geology to meteorology.
3 Letter from Matthew D. Shapiro of 27 January 1987, General Correspondence Box 67, Paul A. Samuelson Papers, David M. Rubenstein Rare Book and Manuscript Library, Duke University.

10 Economics as science
The rules of the game

Economics remains a controversial discipline. Scornfully dismissed in the mid-nineteenth century by Thomas Carlyle as the 'dismal science', it still gets a bad press for its poor predictions or its real or supposed ideological bias. Over time, economists have reflected at great length on their claims to understand reality, and the relationship between their theories and models on the one hand and the social and political reality on the other. They have never been neutral observers interested only in objective knowledge. From Adam Smith to the present day, economists have formulated their theories not merely – in the famous words of Marx – to understand the world but to change it. This makes the distrust that has always been shown towards economists not only understandable but justified. The tension between science and politics is an indissoluble part of the discipline of economics and has repeatedly been the focus of methodological reflection. This applies to John Stuart Mill's distinction between the 'science' and the 'art' of political economy, it applies to Milton Friedman's essay about economics as a positive science and it applies to Lionel Robbins' distinction between pure and applied economics.

This tension has also raised questions as to which instruments are available to the economist for use in substantiating claims to knowledge. Mill sought the answer in the special access the political economist has to his object of study. By means of introspection, he was in a position to study the motives by which people are guided in their economic behaviour, which enabled him to formulate tendencies that, although not precisely quantifiable, were undoubtedly valid. With the development and introduction of new research techniques, this strategy for facing up to the complexity of economic reality has become increasingly implausible. Instead of demanding a separate place for the science of economics beside other sciences, the political economists who came after Mill stressed what they had in common with the rest of science. Hybrid scientists such as Stanley Jevons and Jan Tinbergen transposed techniques of research from their original fields of study in the natural sciences to economics. This was made possible by the assumption that the reality the economist looks at can be described and analysed quantitatively. Just as an astronomer looks for laws in a multiplicity of astronomical observations, so an economist looks for law-like regularities in the multiplicity of statistical facts collected. Only when economists started to see statistical data as the dominant form of empirical material with which they dealt did they develop the kind of instruments that are now used across the board, even in newspapers: price indexes, graphs and tables of accounts. The power of such instruments lies not so much in their quantitative character, as we tend to assume, but in their potential for allowing us to handle complex social reality.

By far the most important instrument developed by economists is the 'model', a set of equations that amount to a description in miniature of an entire economy. For most of the twentieth century, the model was the economists' most important answer to the central question posed by Mill in his essay: how is knowledge of a complex world possible if that complexity cannot itself be controlled? Although admittedly a simplification of the world, a model is at least able not just to reflect complex processes but – if only to a certain level – to enable us to manipulate them.

The next question that concerned economists and philosophers of science, one we still face today, was: where exactly does the power of models lie, in their equations or in their outcomes? Those who stress outcomes seem to be siding with Milton Friedman, for whom the assumptions of economic theories are not relevant. Instead, Friedman pointed to the predictions made using economic theories (models) as the most important criteria by which to judge them. A model is not a 'camera' it is merely an 'engine', a machine for prediction. But we have seen that prediction is only one specific use of modelling. At least as important are simulations carried out using models. The power of model simulations is that they give us 'data' on situations in which we (for example as policymakers) are interested – data we would not otherwise have. Model simulations make demands of the internal and external structure of models. Instead of allowing us to be agnostic about the assumptions contained in a model, as Friedman was, model simulations force us to address questions about the reality content of the model's mechanisms. It is no accident that Tinbergen called his first model a 'mathematical machine'. Faith in the outcomes of model simulations rests to an important degree on faith that the underlying model represents the causal structure of reality.

One thing economists from Mill onwards agreed about was that economics was not an experimental science. Since there were so many 'disturbing causes', economists had no option but to study reality 'in the wild'. They could 'tame' this reality (make it measurable) by means of graphs, price indexes, and mathematics and statistics in general, but they could not actually control it. With the rise of experimental economics, this consensus has been radically overturned. Economists never disputed the possibility of experimenting with people but they did contest the relevance of such experiments for their field of study. Even though behavioural experiments are now a blossoming and fast-growing branch

Economics as science 173

of economics, in my opinion the real breakthrough lies in the introduction of market experiments. A market seems too large a thing to put in a laboratory, but that is exactly what has been happening since the early 1960s.

This breakthrough would not have been possible without the invention of statistical and econometric instruments. In the 1930s, econometrics was described as a combination of statistics, mathematics and economic theory, and the same goes for experimental economics, except that a market is put together not in a computer but in a laboratory. An experimental economist needs to have some idea of the characteristics of the market on which he wishes to experiment. As we have seen, the scaling down and subsequent scaling up of a market is based on characteristics derived from econometric research and expert judgement. This makes such a market experiment a hybrid of model simulation and experiment.

The transformation of his research instruments changed the persona of the economist as a scientist. In the early nineteenth century, an economist was still above all a participant in the public debate, who therefore needed to claim authority for his arguments and judgements. With the introduction of mathematics and statistics in the late nineteenth century, the economist was transformed into a technocrat who developed instruments that, as it were, formed their own independent judgements about the economy. Once the formula for a price index has been determined, the price index itself and not the economist determines whether inflation or deflation is taking place. This does not mean that the economist left the public arena, quite the contrary. Jevons used his price index as an argument in a public debate about the value of gold. Tinbergen's model was a powerful weapon in the public debate about desirable economic policy in the 1930s, as is clear from his participation in and support for the Labour Plan put forward by the Dutch Social Democratic Workers' Party. Over time, a modelling approach turned out to provide far more effective weaponry in the political domain than arguments based on personal judgement and authority. Anyone who wanted to ask questions of the model would have to do so in terms that the model 'understood'. By using models, the economist began to impose impersonal rules of the game on society.

One clear example concerns the quantitative assessment exercises performed by the CPB based on Dutch election manifestos. No other country has a system quite like it. Dutch politicians are expected to behave precisely in the way that Tinbergen described in relation to his own model in the 1930s: they must indicate which variable in which equation needs to be changed. Cuts to state healthcare mean a very different thing if they involve, say, reducing the costs of personnel rather than the maintenance of buildings. The Dutch approach also forces parties to consider the causal connections in the economy. Instead of being a participant in the public debate, the economist is more like an umpire, seeming to stand above the political parties. It is not the economist's verdict that is debated but the choices made by the parties. Although this example is arguably specific to the Netherlands, with its 'polder economy' of consultation and consensus, the eagerness of the economist to adopt the role of objective umpire is not limited to the Dutch situation. Instead of acting in the public domain, the economist increasingly works behind the scenes, out of the public eye, although no more effectively for that. Nowadays, the 'data' that result from a market experiment about a virtual reality can be used as evidence in court.

Anyone who thinks that an economic model (or any other economic instrument, such as an index) can make calculations entirely independently is forgetting the huge amount of human work involved in putting such a model together. Before a question can be put to a model, economists and statisticians have to perform all kinds of tricks to ensure that the model is indeed a description in miniature of an entire economy. The economist's own verdict is now hidden within what looks like an objective representation of an economy. The economist may have lost the heroic status of a Keynes, but his judgement is no less present in his research instruments and the results they are used to produce.

Confidence in these research instruments and the outcomes derived from them is based on the trustworthiness of the experts who create and use them. The consensus between these experts, the economists themselves, is ultimately decisive for the answer to the question of whether an instrument is a good one or not. But the rules of the game behind this consensus are not self-evident, indeed they may not even be in the public interest. Dramatic economic events such as the 2008 financial crisis are cause enough for economists, however reputable, such as Nobel Prize winner Paul Krugman, to call those rules of the game and the accompanying role of the economist into question fundamentally. His choice points to the difference between rigour and relevance. Krugman chose to exchange his role as an engineer and independent umpire for the role that political economists such as John Stuart Mill and John Maynard Keynes played with such verve. They were intellectuals who took part in the public debate by writing opinion pieces and intervening in politics. Krugman's move shows that economists' rules of the game remain contested even within the discipline of economics, especially in times of economic turmoil. Economics is said to be what economists do, but, as economists such as Krugman remind us, it is not enough for economists alone to believe in what they are doing. The rules economists developed over time increasingly gained in rigour, but the relevance of their ideas, methods and practices is, as in the days of Mill, played out in the public sphere.

Further reading

Chapter 1: Introduction

A number of good introductions to the philosophy of science in general and the methodology of economics in particular have been written over the years. *Representing and Intervening* by Ian Hacking (Cambridge University Press, 1983) offers an introduction to the philosophy of science that has become a classic; focusing mainly on the natural sciences, it offers a clear view of the practice of the scientific method. Following on from Thomas Kuhn's seminal *The Structure of Scientific Revolutions* (University of Chicago Press, 1962, new edition 2012), sociologists started to examine scientific practice. This resulted in some highly commendable work. Particularly worthy of mention are K.D. Knorr-Cetina and M. Mulkay (eds) *Science Observed: New Perspectives on the Sociology of Science* (Sage, 1983) and Andrew Pickering (ed.) *The Mangle of Practice: Time, Agency, and Science* (University of Chicago Press, 1995). Bruno Latour pays a good deal of attention to scientific practice in his *Science in Action* (University of Harvard Press, 1987), which although not written as an introduction can be read as one. In the field of economic methodology there is *Spelregels voor economen* by Joop Klant (Stenfert Kroese, 1972), translated into English as *The Rules of the Game: The Logical Structure of Economic Theories* (Cambridge University Press, 1979), as well as *Economic Methodology* by Mark Blaug (Cambridge University Press, 1992), originally published in 1981. Both Klant and Blaug place themselves in the tradition of Karl Popper and Imre Lakatos and attempt to develop normative guidelines for 'good' economic research. In the same tradition is Bruce Caldwell's *Beyond Positivism: Economic Methodology in the Twentieth Century* (Allen & Unwin, 1982). Daniel Hausman's *The Inexact and Separate Science of Economics* (Cambridge University Press, 1992) is firmly focused on analytical philosophy. Wade Hands' *Reflections without Rules* (Cambridge University Press, 2001) is a brilliant synthesis of general philosophy of science and economic methodology, although it focuses rather less on the practice of economic research. Practical research is explored in more detail in the recent *Economic Methodology* by Marcel

176 *Further reading*

Boumans and John Davis (Palgrave, 2010), which attempts to strike a balance between a general introduction to the philosophy of science and an analysis of actual economic research practice. Both Boumans' *How Economists Model the World into Numbers* (Routledge, 2005) and Mary S. Morgan's recent *The World in the Model: How Economists Work and Think* (Cambridge University Press, 2012) combine a pertinent investigation of actual research practice in economics as it has evolved over time with a broad philosophical view, and both are commendable reading for those looking for a deeper methodological analysis of practice. Most recently of all, Julian Reiss has published a comprehensive methodology of economics, written from the perspective of rational choice theory and the analytical tradition in the philosophy of science (*Philosophy of Economics: A Contemporary Introduction* (Routledge, 2013)). Anthologies of original writings by economists on methodology include Bruce Caldwell, *Appraisal and Criticism in Economics: A Book of Readings* (Allen & Unwin, 1984) and Daniel Hausman, *The Philosophy of Economics: An Anthology* (Cambridge University Press, 2008). Anyone wishing to read more about the approach chosen in this book is advised to look at Hans-Jörg Rheinberger's *On Historicizing Epistemology: An Essay* (Stanford University Press, 2010). On the concept of 'persona' in research on the history of science, see 'Scientific Personae and their Histories', *Science in Context* (2003).

Chapter 2: Economics

Whole libraries have been written on John Stuart Mill, although to this day we lack a definitive biography of this central figure of Victorian Britain. Very readable is Richard Reeves' *John Stuart Mill: Victorian Firebrand* (Atlantic, 2007). Mill's economic work is comprehensively analysed from the perspective of modern economic theory in Sam Hollander's two-volume *The Economics of John Stuart Mill* (Blackwell, 1985). From the perspective of analytical philosophy, the book by Daniel Hausman mentioned above is important. Philosopher of science Nancy Cartwright has extended Mill's relevance from economics into the natural sciences with her highly original interpretation of his methodology in *Nature's Capacities and their Measurement* (Cambridge University Press, 1989). An excellent book about the significance of the concept of tendencies for economics today is John Sutton's *Marshall's Tendencies: What Can Economists Know?* (Massachusetts Institute of Technology, 2000).

Chapter 3: Economics and statistics

Four books are available about Stanley Jevons, one of the founders of marginal utility theory. Those that deal most closely with methodological changes to the practice of economics in the Victorian era are the very useful *A World Ruled by Number: William Stanley Jevons and the Rise of Mathematical Economics* by Margaret Schabas (Princeton University Press, 1990) and my own *William Stanley Jevons and the Making of Modern Economics* (Cambridge University Press, 2005). Philip Morowski's controversial but important *More Heat Than Light* (Cambridge University Press, 1989) defends the thesis that the transformation of the method of economics since the 'marginalist revolution' of around 1870 flows from an indiscriminate copying of the methods and theories of natural scientists by economists. Two excellent books about the emergence of statistics in the nineteenth century are Theodore Porter's *The Rise of Statistical Thinking* (Princeton University Press, 1989) and Alain Desrosières' *The Politics of Large Numbers: A History of Statistical Reasoning* (Harvard University Press, 1998). The battle between the Historical School and the Austrian School is treated well in Joop Klant's *The Rules of the Game* (Cambridge University Press, 1979), in Joseph Schumpeter's *History of Economic Analysis* (Allen & Unwin, 1954), still a monument of erudition, and in Mark Blaug's *Economic Theory in Retrospect* (Cambridge University Press, 1962, several editions), which is no less erudite from the perspective of modern economic theory. The best introduction to Lionel Robbins is his own *An Essay on the Nature and Significance of Economic Science* (Macmillan, 1935, several editions). Susan Howson's monumental biography, *Lionel Robbins*, was published in 2011 by Cambridge University Press.

Chapter 4: Business-cycle research

The best book about the rise of econometrics is Mary S. Morgan's *The History of Econometric Ideas* (Cambridge University Press, 1990). *A Case of Limited Physics Transfer* (PhD Thesis, University of Amsterdam, 1992) by Marcel Boumans deals with the work of Jan Tinbergen, paying close attention to his background in the natural sciences and his indebtedness to his teacher Paul Ehrenfest. Boumans' *How Economists Model the World into Numbers* (Routledge, 2005) offers a detailed historical and methodological examination of the rise of modelling in economics. For the development of modelling at the CBS (and later the CPB), see Adrienne van den Bogaard, *Configuring the Economy* (Thela Thesis, 1998). Anyone wanting to know more about the context of statistical research in the Netherlands can turn to *The Statistical Mind in the Netherlands* (Rodopi, 2009), edited by Paul Klep, Jacques van Maarssenveen and Ida Stamhuis. For those able to understand Dutch, other readable works include *Telgen van Tinbergen, het verhaal van Nederlandse economen* by Arjo Klamer and Harry van Dalen (Balans, 1996) and *Verzuilde dromen, 40 jaar SER* by Arjo Klamer (Balans, 1991).

Chapter 5: John Maynard Keynes and Jan Tinbergen

For works on Tinbergen and on the rise of modelling in economic science, see those listed for Chapters 3 and 4. The best biography of Keynes is by Robert Skidelsky, a shorter version of which (although still over 1,000 pages) was published by Palgrave in 2004: *John Maynard Keynes (1883–1946): Economist, Statesman, Philosopher*. Given the huge influence of Keynes' work, there is surprisingly little available on his methodological approaches. The nearest thing we have are Axel Leijonhufvud's *On Keynesian Economics and the Economics of Keynes: A Study in Monetary Theory* (Oxford University Press, 1968) and John Davis' *Keynes's Philosophical Development* (Cambridge University Press, 1994).

Chapter 6: Milton Friedman and the Cowles Commission for Econometric Research

Milton Friedman's 1953 essay prompted an avalanche of publications, mostly in specialist periodicals. A starting point is offered by *The Methodology of Positive Economics: Reflections on the Milton Friedman Legacy*, edited by Uskali Mäki (Cambridge University Press, 2009). Anyone interested in the work of the Cowles Commission should look at Mary S. Morgan's *The History of Econometric Ideas* and at the excellent introduction to the collection of original texts in *The Foundations of Econometric Analysis* (Cambridge University Press, 1997), which she edited along with David F. Hendry.

Chapter 7: Modelling between fact and fiction

The standard account of thought experiments remains Roy A. Sorensen's *Thought Experiments* (Oxford University Press, 1992). Very readable is Tamar Szabó Gendler, *Thought Experiment: On the Powers and Limits of Imaginary Cases* (Garland, 2000), and for those able to read German there is Ulrich Kühne, *Die Methode des Gedanken-experiments* (Suhrkamp, 2005). For an account of thought experiments with a specific focus on economics, see Julian Reiss, 'Causal Inference in the Abstract or Seven Myths about Thought Experiments', *CPNSS Technical Report 03/03* (2002), and for a critical examination of thought experiments in the social sciences, Kathleen V. Wilkes, *Real People: Personal Identity without Thought Experiments* (Clarendon, 1988). An excellent work on Samuelson, with a strongly methodological approach, is Stanley Wong's *The Foundations of Paul Samuelson's Revealed Preference Analysis*, originally published by Cambridge University Press in 1968 and recently reissued. At least as good is Roy Weintraub's *Stabilizing Dynamics* (Cambridge University Press, 1991), which offers a fascinating historical and methodological voyage into Samuelson's main work, *Foundations of Economic Analysis* (Harvard University Press, 1947).

The case study I discuss in this chapter was earlier explored in Daniel Hausman's *Inexact Science* (Cambridge University Press, 1992), but from a quite different perspective. A collection of essays that takes the relationship between fact and fiction as a starting point for asking methodological questions of the science of economics is *Fact and Fiction in Economics: Models, Realism and Social Construction*, edited by Uskali Mäki (Cambridge University Press, 2002).

Chapter 8: Experimentation in economics

Francesco Guala's *The Methodology of Experimental Economics* (Cambridge University Press, 2005) is without doubt the standard reference work on the subject. A particularly interesting read is Ana Cordeiro Santos, *The Social Epistemology of Experimental Economics* (Routledge, 2010). Experimental economists are far more interested in historical and methodological issues than their non-experimental colleagues, which surely has to do with the short history of their method within economics. See, for example, *The Handbook of Experimental Economics* by John Kagel and Alvin Roth (eds) (Princeton University Press, 1995) and *Experimental Economics: Rethinking the Rules* by Nick Bardsley, Robin Cubitt, Graham Loomes, Peter Moffatt, Chris Starmer and Robert Sugden (Princeton University Press, 2010). Philosophers and methodologists of economics, experimental or otherwise, have given much attention to the issue of external validity. Francesco Guala's book covers this topic as well. See also Daniel Steel, *Across the Boundaries: Extrapolation in Biology and Social Science* (Oxford University Press, 2008). Separate mention should be made of a number of penetrating essays by Robert Sugden and Mary S. Morgan, which have appeared in various periodicals including *The Journal of Economic Methodology*. Mary S. Morgan's recent *The World in the Model: How Economists Work and Think* (Cambridge University Press, 2012) contains some of her earlier essays in a much extended and modified form.

Chapter 9: Simulation with models

An extensive literature is available on the subject of model simulations, often looked at in relation to the experiment. I have profited greatly from the following two collections of essays: *Models as Mediators* by Margaret Morrison and Mary S. Morgan (eds) (Cambridge University Press, 1998) and *Science without Laws* by Angela Craeger, Elisabeth Lunbeck and Norton Wise (eds) (Princeton University Press, 2007). See also *Model-Based Reasoning: Science, Technology and Values*, edited by Nancy Nercessian and L. Magnani (Kluwer, 2002), and various articles by Mary S. Morgan on the relationship between model simulations and experiments. There is also an extremely informative special issue of *Science in Context* about simulations, compiled by Sergio Sismondo: *Science in Context* 12(2) (1999). On the functioning of hybrid institutions such as the CPB, see for example the essay collections *Empirical Models and Policy Making: Interactions and Institutions* edited by Mary S. Morgan and Frank den Butter (Routledge, 2000) and *Democratization of Expertise?* by Sabine Maasen and Peter Weingart (eds) (Springer, 2005). On the problems surrounding economic modelling as a non-laboratory endeavour, look out for a forthcoming book by Marcel Boumans, *Clinical Measurement*.

Chapter 10: Economics as science

Most scholarship on the interaction between scientists or social scientists and the public sphere stems from Science and Technology Studies. Only recently has this field begun to address the part played by economics vis-à-vis the public. The classic reference (unrelated to the discipline of economics) is Sheila Jasanoff's *Designs on Nature: Science and Democracy in Europe and the United States* (Princeton University Press, 2005). For nineteenth-century interaction between political economists and the public see Stefan Collini, *Public Moralists: Political Thought and Intellectual Life in Britain 1850–1930* (Cambridge University Press, 1991). On the social sciences in the United States, see Mary Furner, *Advocacy and Objectivity: A Crisis in the Professionalization of American Social Science, 1865–1905* (University Press of Kentucky, 1975) and Mark Solovey and Hamilton Cravens (eds), *Cold War Social Science: Knowledge Production, Liberal Democracy, and Human Nature* (Palgrave Macmillan, 2012). On expertise, including economic expertise, see Harry Collins and Robert Evans, *Rethinking Expertise* (University of Chicago Press, 2007) and Robert Evans, 'Economic Models and Policy Advice', *Science in Context* 12(2) (1999) pp. 351–76. Finally, I should mention Marion Fourcade's important comparison of economic cultures in three countries: *Economists and Societies: Discipline and Profession in the United States, Britain and France, 1890s to 1990s* (Princeton University Press, 2009).

Index

'as if' methodology 77, 91, 94, 96–97, 99, 117, 119, 121, 144, 147, 158
a posteriori method 18
a priori method 18, 21, 25, 33,36
Adelman, Frank L. 152,158–164
Adelman, Irma 152, 158–164
Adelman–Adelman test 152, 157–160, 161, 164, 169
Allais paradox 119, 125–126
altruism 10, 104–105
American Civil War 22, 25–26
analogies between animal and human experiments 122
analytical method 22, 33
Anglican Church 11–13
animal spirits 65
anomaly 125
Antonioni Michelangelo, 1
Apostles, the 63
armchair economics 18, 103–104, 114
Army Corps of Engineers 142
assumptions 6, 77, 81, 89–94, 96–97, 99, 103, 105, 110, 113, 144, 147, 158, 172; domain, heuristic, negligiblilty assumptions 94
astronomical observatory 3
astronomy 5, 16, 20, 99, 150
auctions 124–125, 130–131, 140
Austrian School 21–22, 33–34, 36, 177
average man 20, 30

BAAS (British Association for the Advancement of Science) 11–12, 14
Babbage, Charles 12, 23; his calculating machines 23
Bacon, Francis 12
barometer 23, 39, 56, 67, 76, *see also* business cycle barometer

behavioural experiments 120, 123–126, 146–147, 172
Bentham, Jeremy 9, 13, 22
Berkeley, George 107
Binger, Brian 141
Bloomsbury group 63–64, 71
Blow-Up 1, 5
Bosch Kemper, Jeltje de 40
Bosch Kemper, M.J. de 40, 42–43, 49, 59, 68, 81
Boumans, Marcel 55, 160, 176–177, 180
Bowley, Arthur 40
Brahe, Tycho 79, 98n
Bretton Woods 61, 78
Bridgman, Percy 101, 115
Buchanan, James 149
Bureau of Economic Analysis 3
Burns, Arthur 77–81, 84–85, 88, 90, 95
business-cycle barometer 38–40, 42–43, 49–52, 57; Harvard (ABC) barometer 39
business game 139

Cairnes, John Elliot 21–30, 36–37, 59, 63, 65; *The Slave Power* 22, 25–26, 63
Caltech 116, 127, 135, 137
cameralism 15, 21
Cantillon, Richard 29
Carlyle, Thomas 11, 171
Cartwright, Nancy 18, 176
causation 16, 51, 56, 81, 83, 87–88, 96, 104, 156, 163, 165–166, 172–173, 178, *see also* disturbing causes
CBS (Dutch Central Bureau of Statistics) 3, 38, 40, 42–44, 46–59, 60, 67–68, 75, 81, 156, 170, 177
ceteris paribus conditions 35, 97, 110, 165

Index

Chamberlin, Edward 117, 120, 127–129, 148
Cheysson, Emile 42
Christ, Carl 86–89, 94–95, 98, 150
classroom experiments 120–121, 127–128, 148
Clemenceau, George 64–65
Cliffe Leslie, Thomas Edward 29
common sense 6, 13, 111
complexity 2, 5, 15–20, 30, 33, 51, 65, 71–73, 82, 86, 88–89, 95, 99, 117–118, 122–124, 131, 136, 140, 142, 145–146, 150, 156, 158, 167, 169, 171–172
computer simulations 150, 156–157, 167, *see also* model simulations, model experiments
Conan Doyle, Arthur 26
conceptual analysis 36
conceptual exploration 110–111, 113, *see also* Hausman, Daniel
consensus of experts 167
controlled experiment 2, 16, 61, 89, 93–95, 95, 97, 115, 123, 139, 146–147, 165
correlation 56, 70, 83, 86–87
cost-plus pricing 90–91
counter-cyclical budgetary policy 62, 66 counterexample 92, *see also* testing, hypothesis testing
counterfactual 145–146, 164
Cowles, Alfred 76
Cowles Commission (Foundation) for Econometric Research 69, 76–79, 81, 83–89, 91, 94–99, 150, 161, 178
CPB (Netherlands Bureau for Policy Analysis/Central Planning Bureau) 38, 48, 87, 151, 158, 161–164, 168, 173, 177, 180
credible stories 152
Crome, August Friedrich Wilhelm 31–32
Cuvier, Georges 26

data, statistical data *passim*
deductive method 13–14, 22
detective stories 161
Dickens, Charles 11
dictator game 125
discursive method 22
dismal science 11, 171
disturbing causes 17, 19, 78, 89, 93, 111, 123, 172, *see also* causation
division of labour 108

A Doll's House 67
Domesday Book 15
Domingues, Kathryn 163
Don, Henk 57, 59
dopamine 126
Down Survey 15
Dublin Statistical Society 11–13, 23–24
Duisenberg, Wim 153, 155
Dutch Central Bank 151, 153–158, 161–164, 166–167, 169
Dutch monetarism 153

econometrics 36, 38, 48, 59–60, 67, 69, 76, 82, 136, 156, 162, 173, 177
economic expertise 30, 152, 156, 163, 180
economic methodologists 97
economy(etr)ic modelling 3–4, 6, 38, 62, 76, 88, 155, 166, 168, 172–173, 177, 178, 180
Economic Science Association (ESA) 115
economizing behaviour 34
economy-in-the-model 162
Edgeworth, Francis Ysidro 21, 34–35, 71, 90
Edwards, Ward 103–114 efficiency of markets 126
Ehrenfest, Paul 48, 49, 177
empirical facts 14, 78
empiricism 21–22, 30, 36, 77, 92, 95
Engels, Friedrich 33
Enlightenment 10, 104–105
equilibrium price 127–128, 130
Erasmus University 57, 59
errors 27, 30, 82, 86
Eton 63–64
ECB (European Central Bank) 153
experimental design 120, 124–127, 129–130, 132–133, 136–137, 140, 141, 144, 147, 148, 165
experimental economics 115–116, 122, 124, 127, 147, 149, 172, 173, 179
experimental method 17, 21, 34, 77, 118, 122–124, 142, 149
experimental philosophy 22
experimental science 21, 28, 77, 93–95, 97, 99, 115, 122–123, 172
experimental trading period 129–130, 134, 138, 140, 142–143
experimentation 2, 6, 17, 20, 116–117, 122, 127, 146, 148, 150, 169
experiments of nature 122, *see also* natural experiments

external validity 120, 132–133, 136, 140, 144,147–149, 179
extrapolation 86, 152, 163–164, 166, 167, 179
Eyjafjallajökull 167

Fair, Ray 163
falsificationism 92
Fase, Martin 155, 157
FCC (Federal Communications Commission) 124
Fechner, Ludwig 21
fictional worlds 109, see also thought experiments
financial crisis 62, 96, 167, 168, 174
forecasts 38, 164, 168–169, see also predictions, scenario analysis
Forel, François-Alphonse 42
Foundations of Economic Analysis 100, 178
Fouraker, Lawrence E. 119, 148
framing 126
free trade 15, 25, 31, 33, 66
free will 30
freight transport market 123, 131, 133
French Revolution 9, 11, 104
Friedman, Milton 4, 6, 60, 62, 68, 71, 76–79, 81–100, 110, 113, 114, 115, 116, 117, 118, 119, 120, 122, 146, 148, 150, 152–153, 158, 169, 171–172, 178; 'The Methodology of Positive Economics' (1953) 77, 89, 97, 178
Frisch, Ragnar 48, 81

game theory 117–118, 124, 147
Gelderen, Bob van 41–42
generic experiment 131, 138, see also specific experiment
Gladstone, William Ewart 66
Godwin, William 10, 104–105, 107
Goudriaan, Jan 151
Gradgrind 11
Grether, David 119–120, 148
Guala, Francesco 149, 179

Haavelmo, Trygve Magnus 81
Hall, Robert Lowe 90
Hancock, Neilson 23
Harnett, Donald Lee 119
Harrod, Roy 65
Harvard University Committee for Economic Research 39
Hausman, Daniel 16, 19, 100, 104, 110–111, 113–114, 175–176, 179

Hayek, Friedrich (von) 33
Helmholtz, Hermann 21
Herschel, John 26
Hicks, John 62, 67, 71
historical epistemology 4
Historical School 21, 33, 177
history and economics 4, 15
Hitch, Charles J. 90, 92
Holmes, Sherlock 2, 26
Holtrop, Marius W. 153–157, 162, 167
homo economicus 18
Hong, James (Jim) 121, 129, 131–134, 136–149, 162–165
House of Commons 2, 8, 25
humanities 22, 23, 67, 71
Hume, David 29, 107
Hutcheson, Francis 107
Hutchison, Terence 92–93; *The Significance and Basic Postulates of Economic Theory* (1938) 92
Huygens, Christiaan 31

Ibsen, Henrik 67
ICC (Interstate Commerce Committee) 135
identification problem 82
IMF (International Monetary Fund) 61
incentives 118–119, 143, 147, 165
indicator diagrams 42
induced values 117, 124
induction 12
inductive method 13, 22
inductive statistics 60, 78
inflation 29, 61, 101, 173
initial conditions 92, 93
Institut für Konjunkturforschung 49
instruments, scientific instruments 1, 4–6, 23, 28, 37–38, 63, 77, 82, 93, 95, 100, 104, 171–174; diagrams and graphs as instruments of discovery 27, 31, 51, 130; price index as instrument and argument 29–30, 173
intentionality 34
interventions *in vivo* 164
introspection 16–17, 19, 20, 65, 85, 90, 93, 171
invariance 81–82, 83, 90
invisible hand 101, 103, 108–110, 127
IS–LM model 62, 67

Jesus College, Cambridge 10
Jevons, William Stanley, 19–23, 27–30, 33–34, 36–37, 59, 60, 71, 83, 171, 173, 176–177

Jones, Richard 11–13, 19, 36
Juglar, Clément 40

Kahneman, Daniel 116, 119–120, 123, 125, 147, 148
Kameralwissenschaft 15, 21
Kant, Immanuel 107
Kelvin, Lord (William Thomson) 79
Kennedy, John 100, 110
'Kepler' and 'Newton' stages of science 79
Kessler, G.A. 153
Keynes, John Maynard 3, 6, 28, 37, 40, 60–75, 77, 78, 95, 174, 178; 'Can Lloyd George do it?' (1929) 66; on economics as a 'moral science' 65; as editor of *The Economic Journal* 68; as a literary virtuoso 72; as a moral hero 75; *The General Theory of Employment, Interest and Money* (1936) 62, 65–67, 71; *The Economic Consequences of the Peace* (1919) 63–65, 73
Keynes, John Neville 63
Keynesian interventionism 99
Klant, Joop 3–4, 175, 177
Klein, Lawrence 77, 85–89, 118, 150, 152, 158, 159, 160, 163
Klein–Goldberger model 158, 160, 163
Knies, Karl 33
Koninklijke Vereniging voor de Staathuishoudkunde (Dutch Royal Economic Society) 40, 48, 151
Koopmans, Tjalling 76, 79, 81, 84–85, 88, 90, 98
Krugman, Paul 3, 174
Kuhn, Thomas 98, 107, 111–112, 125, 175

laboratory, economic laboratory 2, 17, 21, 23, 27, 34, 93, 94, 110, 116–118, 121–123, 125, 127, 131–133, 136, 140, 142–147, 149, 163–166, 173, 180
laboratory representations 133
labour theory of value 20
Lange, Oskar 69–72
Latour, Bruno, 4
law-like regularities 16, 123, *see also* tendency laws
League of Nations 62, 67–69, 77, 79, 87, 155
Leontief, Wassili 100
liberalism 8, 30
limnimeter 41–43

Lipsey, Richard 123
liquidity preference 62
List, Friedrich 31
Lloyd George, David 64–66
Logic of Scientific Discovery (1956) 92, *see also* Popper, Karl Raimund
London School of Economics 40
Longfield, Mountiford 24
looking 'under the hood' 98
Lopokova, Lydia 67
Lucas critique 150
Lucas, Robert 62, 150

MacKenzie, Donald 4
Magdalen College, Oxford 13
Malthus, Thomas Robert 9, 10–12, 104–105, 107, 114; *Essay on the Principle of Population* (1798) 10, 104, 114
Malthusian population question 64
Manchesterthum 33
Mandeville, Bernard 107–108, 112; *The Fable of the Bees* 107, 112
Marey, Jules 41–42
market efficiency 99
market forces 101
market simulations 139, *see also* market experiments
market transparency 131
markets as calculators 121
Marschak, Jacob 69–72, 76, 79
markets and agents 'in the wild' 118, 124, 132–133, 144, 172
markets and institutions 109, 111, 113, 121, 126, 129, 131–132, 147
Marshall, Alfred 4, 71, 95, 176
Marx, Karl 9, 171
means-ends rationality 34
measurement without theory 77
mechanics 16, 26
Menger, Carl 21, 33
mental experiments 17, *see also* introspection
meteorology 5, 20, 23, 38, 39, 150, 170
methodology 4–6, 14, 20, 76–77, 89, 92, 99, 116, 175–176; normative 4, 13, 19, 65, 96, 97, 99, 175; prescriptive 4
Methorst, Henri 40
micro-economic theory 20, 103
Mill, James 8, 9
Mill, John Stuart 2–3, 5–6, 8–9, 11, 13–28, 31, 35, 37, 59, 62–63, 65, 77–78, 89, 90, 95, 98–99, 111, 122–124, 146, 171–172, 174, 176;

A System of Logic (1843) 8, 14; his autobiography 9; his essay on the definition and method of political economy (1836) 5–6, 8–9, 14, 18, 77, 89, 98, 123, 172; *Principles of Political Economy* (1848) 18, 23, 25; the special character of human dealings 18
Miller, Merton 97
miniaturized markets 122
Mises, Ludwig von 33–34
MIT (Massachusetts Institute of Technology) 108, 115, 155
Mitchell, Wesley Clair 77–81, 84–85, 87–88, 90, 95, 96
model experiments, model simulations 152, 157–158, 161, 164, 165–167, 169, 172, 179
models, economic models *passim*; as explanatory or predictive instrument 52, 86, 152, 168; their inner coherence and structure 162; as neutral policy platform 57; their plausibility 71, 105, 107, 146, 167, 168; as representations of the structure of the economy 14, 29, 47, 58, 78, 88, 119, 122, 131, 138, 140, 143, 147, 158, 161, 166, 173; as scientific and policy instrument 4–5, 48–49, 55–57, 150–151, 153, 155–156, 158, 161–162, 166–169, 172–173, 180
Modigliani, Franco 155
monopolistic competition 117
Moore, Henry L. 83–84
moral perfection 63
moral sciences 8, 14, 16–18
Morgan, Mary S. 40, 72, 75, 82–83, 152, 176–180
MORKMON (MOnetair en Reëel Kwartaal MOdel van Nederland) 152

'naive' models 86–88, 95
Napoleon 11, 15
Napoleonic Wars 9
NSF (National Science Foundation) 116
natural experiment 122
natural sciences 2, 16–18, 20–23, 33, 35, 59, 77, 79, 93, 105, 111, 171, 175–177
natural scientist 2–3, 16, 28, 30, 33
Nature 3, 22, 37, 82, 115, 176, 177
NBER (National Bureau for Economic Research) 68, 76–78, 95, 158

neoliberalism 99
neuroscientists 115
neutrality of economics 13, 43
Nixon, Richard 61, 78
Nobel Prize in Economics 3, 48, 62, 85, 99, 103, 116, 123, 149, 150, 174
noise, white noise 82, 159, 161, 164
non-laboratory science, economics as a 119
normative impulse analysis 153
note-taking 23, 25, 75

Obama, Barack 126
Obamacare 126
objective knowledge 171
Office for National Statistics 3
oligopoly theory 118
operationalism 101, 115
opportunity costs 5
Oreskes, Naomi 143
Oriel College, Oxford 13
Oxford Research Group 90
oxytocin 126

paracetamol 17
paradox 100, 103, 105, 107–108, 119, 125, 126, *see also* Allais paradox
Pearson, Karl 70
perfect observer, qualities of the 26
perpetuum mobile 105
persona of the economist 3–7, 173, 176; as detective 25, 30, 59, 154, 162; as a hybrid between natural and social scientist 3; as a hybrid expert 5, 47, 49, 59; as instrument maker 7, 28, 30, 38; as model-builder 61; as playwright 25, 30, 59; as a theoretician 100, 115
Persons, W.M. 39, 83
Petty, William 15
philosophy of science 4, 6, 61, 92, 93, 94, 148, 175, 176
physique sociale 20, *see also* social statistics
Pickering, Andrew xii, 114, 175
Plan van de Arbeid (Social Democratic Workers' Party Labour Plan) 48
plausible stories versus testable hypotheses 71
Plott, Charles (Charlie) 116–124, 127, 129, 130–134, 136–149, 162–165
Polak, J.J. 57–58, 82, 155
policy scenarios 48, 59, 151, 158
political arithmeticians 15

186 Index

Political Economy Club 14, 25
Poor Laws 9, 11, 19
Popper, Karl Raimund 92–94, 97, 175; *Logik der Forschung* (1934) 92
posted-prices 121, 129–132, 134–135, 137–140, 142, 146
predictions 38, 85–88, 91, 93–94, 96, 99, 113–114, 118, 125–126, 150–152, 158, 164,168–169, 171
probability theory 20, 70, 82–83, 96, 122, 125, 147
profit maximization 90–91, 96
protectionism 15, 31
psychological laws 17
psychophysics 21
public debate 3, 9, 25, 26, 59, 66, 173–174
public goods 126
public intellectual 3, 10, 63, 70, 174, 180
public interest 56, 107, 174
public policy 7, 116, 155, 161

quantity theory of money 29
Queen Elizabeth I 9
questionnaires 85, 90, 91, 146
Quételet, Adolphe 3, 12, 20, 23

railway timetables 151, 158, 169, *see also* policy scenarios
RAND corporation 90
rate filing policies 132
rational choice theory 96, 119–121, 125–126, 149, 176
reference cycle 78–80
replication and reproduction of experiments 130
reservation price 128
Revue des Deux Mondes 24
Ricardian economics 14, 19
Ricardians 14
Ricardo, David 2, 9–18, 36, 62, 65
Robbins, Lionel 21–22, 34, 35–37, 61, 65, 90, 92, 100, 114, 123, 171, 177
Roscher, Wilhelm 33
Rousseau, Jean-Jacques 104–105
Royal Economic Society 40
Royal Statistical Society *see* Statistical Society of London
Rutgers University 78

Samuelson, Paul 6, 100–104, 108–111, 113–115, 118, 122–123, 148–149, 152, 164, 178; 'An Exact Consumption-Loan Model with or without the Social Contrivance of Money' (*JPE* 1958) 100; his Harvard thesis 'Foundations of Analytical Economics: The Observational Significance of Economic Theory' 100; scorning the (empirical) 'hard work' of the social scientist 104, 111; his textbook *Economics* (1948) 108
Savage, Leonard Jimmie 96, 119, 125
scale models 142–143
scaled-down markets 121, 132–133, 136, 140, 143
scaling down time 142–143, 163
scaling up 132–133, 138, 145, 163
scenario analysis 49, 158, 168
Schultz, Henry 148
Schumpeter, Joseph 33, 100, 177
Scrooge 11
Section F of the BAAS 12, 14
self-interest 10, 12–13, 26, 103–104, 108, 112
selfish markets 109, 111
Senior, Nassau 11, 13, 18
Shapiro, Matthew 163
shifting the burden of proof 121, 146
shocking a model 87, 159–162, 164, 169
Siegel, Sidney 119–120, 127, 148–149
Simon, Herbert 103, 106–107
simulation 59, 120, 123, 139–141, 150–152, 159–161, 163, 166–167, 169, 173
Skidelsky, Robert 63–64, 67, 71, 75, 178
small-scale worlds 136
Smith, Adam 9, 11, 15, 31, 101, 107–108, 112, 171; *The Wealth of Nations* (1776) 107
Smith, Vernon 116, 123–124, 126–129, 139, 147–149
Snow, C.P. 70–72
Snow, John W. 132, 149
social physics 20
social sciences 5, 16, 20, 76, 79, 82, 178
social scientist 3, 5, 17
social statistics 3, 20, 23, 40
Sombart, Werner 33
specific experiments 132, *see also* generic experiments
St John's College, Cambridge 11
statistical bureaus 3, 40, 55, 158
statistical facts 3, 15, 64, 172, *see also* data, statistical data
Statistical Society of London 12, 14
statistics *passim*

statists 15, *see also* political arithmeticians
steam engine 16, 42
Stevin, Simon 105
Stigler, George 95, 97
strategic decision-making 124, 147
stress tests 167
structural equations 82, 90
structural models 81–83, 86, 88, 150, 155, 169; as a representation of the causal structure of an economy 83
subjective probability theory 70
Svorencik, Andrej 127, 135, 141, 149

target market of an experiment 134
Taylor, Harriet 8
techniques of the economist 2–4
television economists 3
tendency laws 13, 18
test bed experiments 165
testing, hypothesis 38, 60, 68, 71–72, 82, 91–92, 99–100, 103–104, 110, 146, 163
the market in the experiment 118
the photographer's darkroom and the work of the economist 1–2, 104
thinking in numbers 23
thought experiments 6, 10, 17, 105, 100, 104–105, 111–112, 107–108, 112–115, 122, 152, 178; compared to a trip to a virtual world 108
Thurstone, Louis Leon 116, 148
Tinbergen, Jan 38, 43, 47–49, 51–73, 75, 77–79, 81–83, 87, 88, 90, 98, 118, 147, 151, 152, 155, 158, 161, 169, 171–173, 177, 178
Tocqueville, Alexis de 33, 37
Torrens, Robert 14
trading situation in experiments 141
Trinity College, Cambridge 12
Trinity College, Dublin 14
Turing test 160
Tversky, Amos 119–120, 125, 147, 148

UMTS (Universal Mobile Telecommunications System) 124
University of Arizona 116
University of Chicago, 69, 78, 175
University College, London 22, 26

University of Pennsylvania 155
US Department of Transportation 132, 149
US Treasury Bills 130
utilitarianism 9
utility theory 20–21, 33, 125, 148, 176; marginal utility theory 20–21
utopia 104–105, 107
Uyl, Joop den 58

Verbruggen, Johan 57, 59, 168
Versailles Treaty 63–66, 75
Victorian England 8, 9
Victorian periodical culture 9
Vietnam War 61
virtual experiment 152

Wagemann, Ernst, 49
Walras, Léon 33, 85
Watt, James 42
Weber, Max 33, 34
Weisberg, Michael 142
Westminster Review 2, 8, 9, 14
'what if' stories 151, *see also* counterfactuals
Whately, Richard 11, 13, 18, 108, 111
Whewell, William 11–16, 18–19, 25, 28, 98; coined the term 'scientist' 11; *History of the Inductive Sciences* (1837) 11; *Philosophy of the Inductive Sciences* (1840) 11, 28
Wibaut, F.M. 47, 49, 59
William of Orange (the Silent) 105
William the Conqueror 15
Williams, Fred E. 129, 130, 131, 140
Wilson, Woodrow 64–65
Witt, Johan de 31
Wolff, Sam de 42, 68
'Wonder is no Wonder' 107, *see also* thought experiment
Woolf, Virginia 64, 67, 71
Wordsworth, William 11, 19
World Bank 61
Wundt, Wilhelm 21

Yale University 69, 76, 79, 164

Zijlstra, Jelle 155